Introduction to

DIRECTIONAL AND HORIZONTAL DRILLING

Introduction to

Directional and Horizontal Drilling

J. A. "Jim" Short

PennWell Books

PennWell Publishing Company
Tulsa, Oklahoma

DISCLAIMER

This text contains statements, descriptions, procedures, and other information, hereinafter collectively called "contents," that have been carefully considered and prepared as general information. The contents are believed to represent situations and conditions reliably that have occurred or could occur but are not represented or guaranteed as to their accuracy or application in any condition or situation. There are many variable conditions in oilwell and gaswell drilling and related situations, and the author has no knowledge or control of their interpretation. The contents are intended to supplement and not to replace the user's judgment in considering, investigating, and verifying actions and situations. Use of the contents is solely at the risk of the user. In consideration of these premises, any user of the contents agrees to indemnify and save harmless the author from all claims and actions for losses and damages.

PennWell Publishing Company
1421 South Sheridan/P.O. Box 1260
Tulsa, Oklahoma 74101

Library of Congress cataloging in publication data

Short, J. A.
Introduction to directional and horizontal drilling / J.A. "Jim" Short,
p. cm.
Includes bibliographical references and index.
ISBN 0-87814-395-5
1. Directional drilling. 2. Horizontal oil well drilling. I. Title. II. Title: Directional and horizontal drilling.
TN871.23.S48 1993
6221,.3381--dc20

93-16840
CIP

Printed in the United States of America

1 2 3 4 5 97 96 95 94 93

**This book is dedicated to my wife,
Catherine Leona "Campbell" Short.
She has enriched my life, continually
reinforcing our relationship over the
years. She truly personifies the generous,
loving wife and mother.**

Miss Kitty, I love you.

CONTENTS

This book should raise as many questions as you might have had before you started reading it . . . maybe more. That's not meant as an apology, but as a challenge!

—William L. Leffler (Petroleum Refining for the Nontechnical Person, Second Edition, 1985, PennWell Books).

PREFACE

Vertical drilling is fundamental to the oil and gas industry. Directional drilling developed from a need to vary direction from vertical drilling and has been facilitated by advances in technology. It is a commonly used, well-established, and proven technique. Horizontal drilling developed for similar reasons. It is widely used and is gaining acceptance in the industry. Through continued use and technological advances, additional applications of these two innovative drilling methods will develop, further increasing their importance. Both are used worldwide to prevent waste by developing and producing oil and gas not recoverable by other methods and by reducing costs.

This book is an introductory text on directional and horizontal drilling and related activities. The material is presented in non-technical language with explanations of common terminology. The text follows the natural sequence of events; new subjects build upon prior material in a building-block fashion. This serves a dual purpose. Those less-experienced can start at the beginning, laying a foundation and building upon it. More advanced readers may go directly to subjects of interest. Each chapter starts with a summary for a quick review and ends with a comprehensive list of references as sources of additional information. Specific topics can be found easily from the Table of Contents or in the expanded Index.

This book is for anyone interested in directional and horizontal drilling. It should be very helpful to beginning employees as well as to personnel in other sectors of the oil and gas industry, including those in related fields such as service and supply companies. Read the book to learn general information about directional and horizontal drilling, scan it for special subjects, or use it as a reference or textbook.

CHAPTER 1

OVERVIEW, DESIGN GUIDELINES

SUMMARY

By earlier methods, all wells were drilled vertically downward. Directional drilling evolved from the need to drill the hole in other directions. Special drilling tools and procedures are used to change the direction of the wellbore from vertical to directional or horizontal in order to penetrate targets that cannot be reached by regular vertical drilling methods. Directional and vertical drilling serve mainly for the drilling of exploration and development wells. Horizontal drilling creates development wells with increased, sometimes very high, production rates. There are various well patterns within the directional and horizontal classifications, depending upon the type of well.

Directional and horizontal drilling are high-risk drilling operations compared to vertical drilling. Efficient drilling programs must be designed carefully. Successful designs have a drillable well path, provision for casing, and minimized hole problems. The well path includes the kickoff depth, the angle-build and angle-drop rates, the drift and direction of the wellbore, the target, and limits.

Directional and horizontal drilling are flexible and applicable to many situations; these wells are drilled worldwide in most major oil and gas fields, both on land and offshore. Usage is increasing, with a potential for widespread future usage.

HISTORY AND DEVELOPMENT

. . . they may be a witness unto me that I [Abraham] have digged this well. Genesis 21:30

The history of drilling fades into the distant past. China had wells before 1200 A.D., later followed by drilling in France, Italy, and West Virginia. The first drilling objective was to produce water. Later needs for resources led to drilling steam for geothermal energy, saltwater for salt, and gas for heating and oil. The Drake well, drilled in Pennsylvania in 1859, is the acknowledged start of the drilling industry in the United States. Drilling equipment began with hand-digging tools, followed by spring pole, cable-tool, and rotary rig equipment in the late 1800s. Early "churn" drilling used a cable or flexible drilling line so that holes were mainly vertical.

Rotary drilling with a rotating drillstring developed into a highly efficient process for drilling and completing oil and gas wells at depths greater than 30,000 feet. Rotary rigs drill on land or offshore, and some are modified for special drilling services. Rotary drilling methods were later modified for directional drilling.

Directional tools and techniques evolved slowly from vertical drilling. An early reason for directional drilling was due to a "fish," unrecoverable drilling tools lost in the hole. Directional methods allowed drilling around and bypassing the fish, a less expensive option than drilling another hole. Crooked holes were another problem that led to directional drilling. One other potential and less publicized incentive may have been to drill into more productive areas under adjacent acreage where ownership may have been in question.

The whipstock was the first reliable directional drilling tool. Development of new tools and techniques aided first in drilling straight and vertical holes and later aided directional drilling. Developments in measuring instruments were the final step leading to modern directional drilling.

Directional drilling is conventionally defined as a procedure for drilling a nonvertical hole through the earth. It first gained prominence when it was used to control a blowout well in southeast Texas in the mid-1930s. At a safe distance from the blowout, a directional hole was drilled at an angle to a point near the bottom of the blowout hole. Fluid was pumped through the deviated hole into the formation, stopping the blowout. This innovative procedure done on a sensational and highly productive well received widespread publicity. It focused attention on the somewhat new drilling procedure.

Directional drilling had a strong start offshore and in other areas where it was difficult or expensive to build a surface location. Early offshore wells were drilled on wide spacing from piers and later from individual platforms. Directional techniques allowed drilling multiple wells from one location, thus eliminating construction of an expensive structure for each well (see Fig. 1–1). These and similar procedures firmly established directional drilling, and it developed into a reliable, efficient drilling procedure with widespread usage. (Note that the angles of bends are exaggerated in most illustrations to allow easy visualization.)

As the drilling industry has matured, wells have been drilled vertically to more than 30,000 ft deep. However, very deep drilling has become less common because of the expense and indications that oil and gas do not often occur at these depths. This, in part, has led to extended-reach, drilling directional to greater distances.

Horizontal drilling subsequently evolved mainly to improve well productivity. It involves drilling the well in a curve from vertical to horizontal and then horizontally. The first wells had one or more short holes drilled horizontally into the formation from the vertical wellbore. These "drain holes" exposed more of the reservoir to the wellbore and produced larger volumes of oil and gas.

The horizontal drilling procedure had been tested in various countries by the 1950s. However, inadequate equipment, lack of demand, and the relatively high cost compared to conventional recovery techniques hampered development. Interest revived in

Figure 1–1
Multiple wells drilled from one location

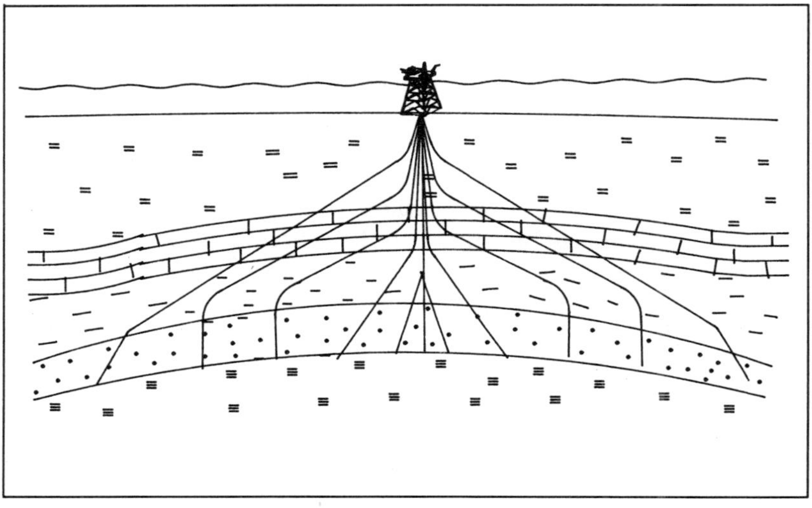

the 1980s, focusing on drilling a single hole a longer horizontal distance into the formation. Tools and techniques developed at an accelerated rate, further increasing efficiency. Horizontal drilling has many applications. It is the latest (and very significant) drilling technique.

DIRECTIONAL STATUS AND APPLICATIONS

Modern directional drilling is an established, widely used drilling procedure. It was originally developed for sidetracking a fish, drilling kill wells, correcting crooked-hole problems, and later preventing the well from crossing lease lines (see Fig. 1–2). It is still used for these purposes. They are important, but other equally important applications have developed over time, such as drilling for attic oil and gas. Directional drilling is common in both offshore and land operations. Major areas of usage include the Texas-Louisiana Gulf Coast, the North Sea, the Mideast, and the Far East.

Equipment and techniques permit drilling any reasonably designed well pattern. Regular directional patterns are more common, with slant and extended-reach holes drilled where applicable. Directional patterns can be combined with horizontal patterns, and expanded usage will lead to other applications.

MULTIPLE WELLS FROM ONE SURFACE LOCATION

Drilling multiple wells directionally from one surface location is a common, important application of directional drilling. Multiwell drilling sites include offshore platforms, man-made islands and peninsulas, and platform and earthen locations in swamps, jungles, and other isolated areas. The older, highly developed East Wilmington field in California is a significant example of a multiwell site. It has nearly 1,200 wells, including high-angle and extended-reach, from 4 man-made islands and 4 earth-filled pier locations.

Modern directional and extended-reach techniques may drill into large areas containing oil and gas from one surface location (see Fig. 1–3). A vertical well penetrates the reservoir at one point. Directional drilling increases coverage substantially as illustrated by the following, based on about 15,000 ft of deviated hole.

Holes at 20° cover about 3 square miles. Coverage increases about 340% at a low inclination angle of 40°. Increasing the angle

Figure 1–2
Early directional applications

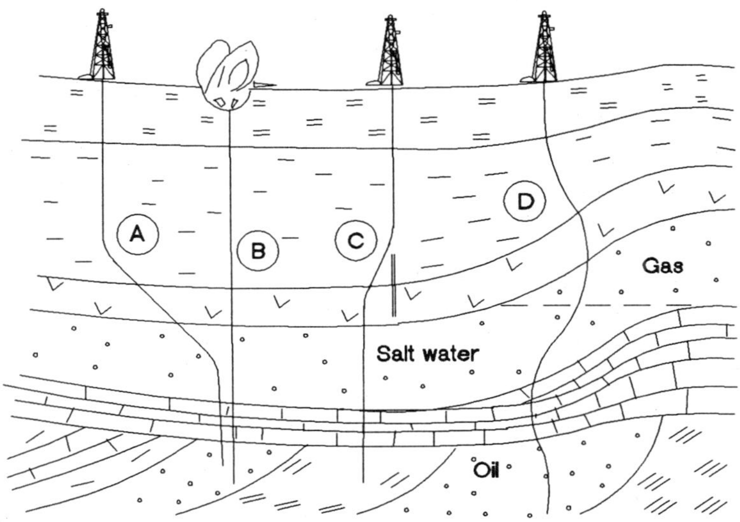

to 60° increases coverage about 200% more than that at 40°. High-angle extended-reach drilling at 80° increases coverage about 130% more than that at 60°, 234% more than coverage at 40° and 820% more than coverage at 20°. A significant example of this is an offshore well in Australian waters drilled to a measured depth of more than 18,000 ft. Horizontal displacement was almost 3 miles at a true vertical depth of less than 8,000 ft. About 28 square miles of reservoir were theoretically accessible to one surface location in this extreme case. This area is considerably larger than the average size of most oil and gas fields.

There are various advantages to drilling multiple directional wells from the same surface site. The main advantage is the single site requirement. It is more economical to drill many directional wells from one platform than it is to build a costly platform for each vertical well. The same situation occurs in swamps, jungles, and

Figure 1–3
Directional wells increase coverage

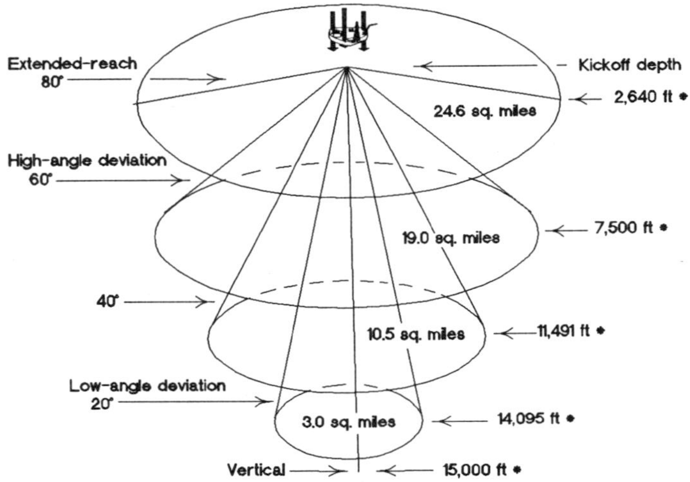

Based on drilling 15,000 ft, measured depth, of deviated hole below the kickoff point

* = True vertical depth below kickoff point

other isolated areas because of the costs of building access roads and multiple surface locations. Common gathering, separation, storage, and other production facilities further reduce costs.

Many productive formations do not contain sufficient volumes of oil and gas to justify the costs of building individual platforms or single-well locations in order to drill vertical wells. The more cost-effective procedure of drilling multiple wells from a single location often allows economical development and production. This allows production of oil and gas that would not otherwise be produced.

INACCESSIBLE SURFACE LOCATIONS

Inaccessible surface locations inhibit development by the drilling of individual vertical wells for various reasons. Some surface locations are inaccessible for economical, physical, or other reasons. Surface drilling sites are very costly, if available, in residential and industrial areas. Ordinances and statutes prevent drilling in some areas. Shipping fairways must be left open for ships to pass, so a drilling platform cannot be constructed on the fairway. Other restricted areas include parks, lakes, cemeteries, recreational areas, and major thoroughfares. Related reasons for not drilling in some areas include concerns about safety, noise pollution, and the difficulty of maintaining long-term production and transportation facilities. The only reasonable method of recovering the underlying oil and gas in these situations is by directional drilling. It often is possible to obtain a few acres for a single surface drill site and then drill multiple directional wells into the surrounding area from the single site.

CHANGED AND MULTIPLE TARGETS

Many wells are nonproductive dry holes. Geological and reservoir information obtained during drilling may suggest a productive

Figure 1–4
Other directional well applications

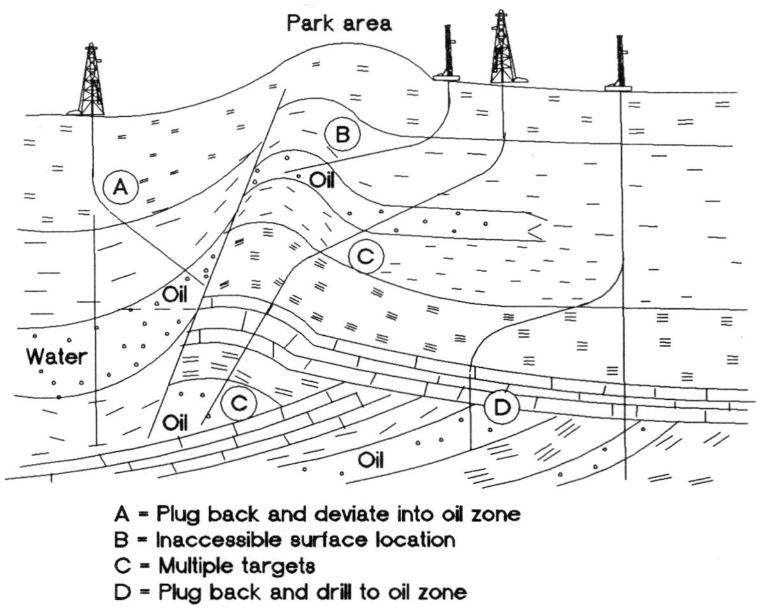

Figure 1–5
Salt dome drilling

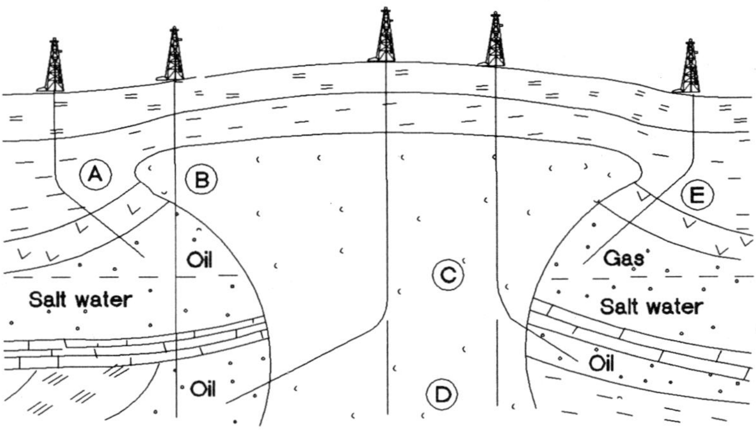

A = Attic oil B = Dual completion C = Sidetracks
E = Original dry holes F = Attic gas

area near the wellbore. It is common in this case to plug back, sidetrack, and drill directionally into the productive area. Oil and gas frequently overlay water in dipping reservoirs. A vertical hole drilled into the water zone may be sidetracked for drilling directionally updip into the oil and gas zone. A well may be drilled directionally under an inaccessible location. Wells can be drilled directionally into multiple targets for dual completions (see Fig. 1–4). Similarly, an oil well in the gas cap or a dry hole may be sidetracked and drilled into the underlying oil zone. Basement oil, attic oil and gas, and salt dome and fault traps are common directional drilling targets (see Fig. 1–5).

Exploration wells may be drilled directionally from a single location in a similar manner. Normally, exploration wells are drilled vertically and the field is developed with directional wells, generally from a single surface location such as a platform. Sometimes the exploration prospect may require multiple exploration wells, and the cost of individual surface locations is very expensive. Then a single surface location is built, such as an ice island in arctic waters. Regular and long, extended-reach exploration wells may be drilled for exploration and later developed if justified.

Drilling into multiple targets is another directional drilling procedure. Oil- and gas-bearing strata may occur at different depths and horizontal locations in a localized area. These may be tested and produced by deviating and drilling directionally into these multiple targets with a single directional well under favorable conditions.

SLANT HOLES

Slant holes are a special application of directional drilling in areas where strata containing oil and gas occur at shallow depths. They are similar to drilling multiple directional wells from a single surface location with several differences. In these cases, the vertical distance to the reservoir is too short to establish sufficient curvature and drill directionally into targets a long horizontal distance from the wellbore. The drilling starts from the surface at an angle of 30°–45° with a slant hole rig. The bottom of these holes may be displaced over 5,000 ft horizontally at vertical depths of 3,000 ft (see Fig. 1–6). This is about twice the horizontal distance obtainable with conventional directional drilling to the same depth. Otherwise, slant hole drilling serves the same purpose as extended-reach directional drilling and has similar advantages. Some

Figure 1–6
Slant hole and slant/horizontal combination

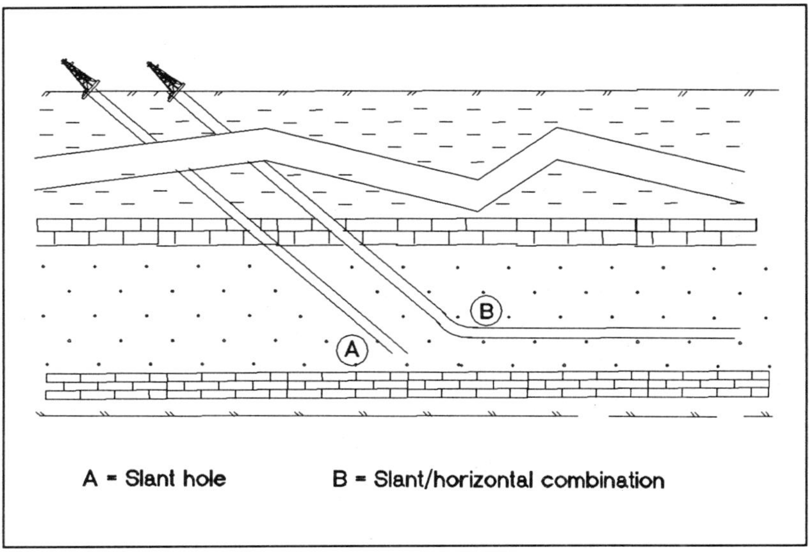

areas of slant hole drilling include Canadian gas sands, Peruvian offshore waters, the Far East, and the Athabasca heavy oil sands in Canada.

HORIZONTAL STATUS AND APPLICATIONS

Horizontal drilling is a procedure for drilling and completing oil and gas wells with improved productivity compared to wells drilled by other methods. A curved section is drilled from the bottom of the vertical hole, followed by drilling horizontally into the formation. Horizontal drilling may be combined with other forms of directional drilling, such as a horizontal section at the bottom of an extended-reach well. Horizontal drilling is well established, adaptable to a wide range of situations both on land and offshore, and its usage is growing rapidly.

Most major fields have horizontal and some combination wells. General areas of activity include Canada, Indonesia, France, Africa, the North Sea, and Mideastern countries such as Saudi Arabia. The highest level of activity is in the United States. Some states, such as Texas, have statutes governing aspects of horizontal drilling such as well spacing and production schedules.

A field or reservoir may require fewer horizontal wells for complete development as compared to other methods of drilling. Vertical or directional wells efficiently deplete or drain a given area of reservoir. Horizontal wells increase the area of drainage by a multiple related to the length of the horizontal section, which is generally considerably more than the average vertical or directional well. The net result is fewer horizontal wells for developing a given size field as compared to vertical and directional wells. Directional and extended-reach drilling increase areal coverage from one surface site, and combining these with horizontal drilling further reduces the number of wells needed.

INCREASED PRODUCTIVITY

Horizontal wells have higher production rates and produce greater quantities of oil and gas than wells drilled by other methods, as verified by production histories and computer simulations. The common contact surface area between the wellbore and the formation limits the flow of oil and gas into the wellbore. Production is roughly proportional to the reservoir area contacted. Horizontal wells have long holes drilled horizontally into the formation compared to shorter sections in vertical and directional wells. The net result is that the wellbore and formation have a larger common open section, thus allowing larger volumes of oil and gas to be produced. The situation is analogous to draining water out of a water tank with a large diameter pipe compared to a small diameter pipe.

Reservoir flow mechanics define the flow of oil and gas in the reservoir. According to the radial flow theory, oil and gas flow radially inward toward vertical and directional wellbores. The cross-sectional area available for flow decreases as oil and gas approaches the vertical wellbore. This increasing flow restriction uses more reservoir energy to produce a given amount of oil and gas.

However, linear flow theory has more influence on flow into horizontal holes, at least near the wellbore and during the early producing life. Flow mechanisms are complex and reservoir fluids have a fixed amount of energy. In summary, higher energy requirements restrict the flow rate from vertical and directional wellbores more, compared to the lower energy usage and correspondingly larger flow rates from horizontal wellbores. This more efficient use of energy also enhances total recovery from the well before it reaches the economic limit for production.

Horizontal drilling also improves productivity from low-permeability formations. Many formations contain oil and gas but produce low volumes from vertical and directional wells because of low permeability. Horizontal wells have increased flow rates because of the increased flow area and decreased reservoir energy requirement as described. Therefore, many low-permeability formations are noncommercial with vertical and horizontal drilling but produce economic volumes of oil and gas from horizontal holes. Because of their greater exposure to the producing zone, horizontal wells also may be more effectively hydraulically fractured (creating multiple fractures compared to a few fractures), which further increases productivity (see Fig. 1–7).

Oil and gas often occur in thin formations. Small volumes of oil and gas near the wellbore, sometimes combined with low-permeability, may further restrict flow rates. Long horizontal sections increase flow rates as described for other situations.

There are many examples of increased productivity from horizontal holes. A horizontal well in the North Sea flowed 30,000 BOPD, approximately 10 times the production rate of an average vertical or directional well in the field. The Austin Chalk formation in southern Texas has many horizontal wells. The average for 15 wells with various horizontal section lengths was 460 BOPD and

Figure 1–7
Horizontal wells and low permeability

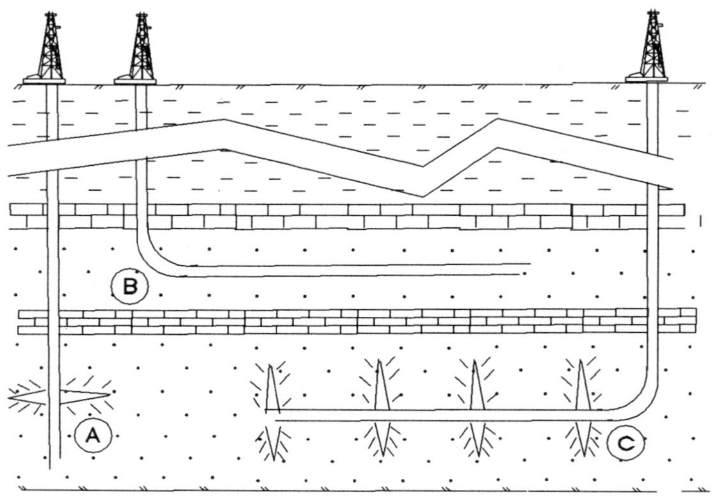

A = Vertical well, single hydraulic fracture
B = Increased wellbore exposure to formation
C = Multiple hydraulic fractures

260,000 cubic feet of natural gas per day (260 Mcfd). This is about 3 to 5 times the amount of production from an average vertical or directional well.

VERTICAL FRACTURES

Vertical, or highly tilted, natural fractures frequently contain oil and gas. These may cover wide vertical areas and contain large volumes. Sometimes oil and gas may flow slowly into the fractures from adjacent low-permeability formations, effectively recharging the fractures. A vertical or directional well may penetrate one fracture but seldom more than two. Often several fractures must be penetrated for the well to be economical. A horizontal well frequently penetrates several fractures (see Fig. 1–8). Steeply dipping productive formations can be a comparable situation.

A significant example of a field with high-angle or vertical fractures is the Pearsall field in south central Texas. An average vertical well produces about 30,000 bbls during its lifetime. This is uneconomical. Some horizontal wells have already exceeded 100,000 bbls. One well produced more than 100,000 bbls in 16 months, and the projected ultimate recovery is 375,000 bbls. This suggests recoveries from horizontal completions will be at least 3 and possibly 5 times that of vertical wells. As a note of caution, there are older vertical wells that would not be commercial even with these increases.

Analogous situations are isolated areas of high-permeability containing oil and gas. These include sand lenses and dune-type features isolated within a dense or low-permeability formation (see Fig. 1–8). Vertical or directional wells commonly drill into only one of these high-permeability areas, and the flow rate may not be economical. A horizontal well can drill through several of these to produce at a higher and often economical rate. A well in the North

Figure 1–8
Other horizontal well applications

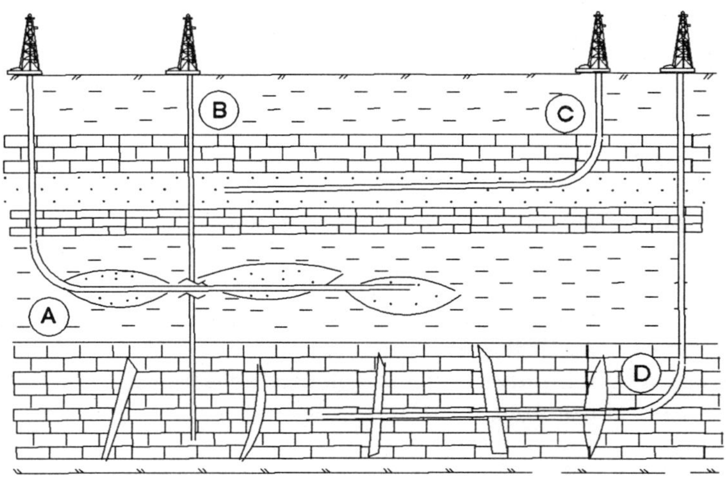

A - Multiple sand lenses
B - Vertical dry hole
C - Thin zone
D - Fractured formation

Sea area drilled a 2,000-foot horizontal section and encountered several good dune-type features. Initial maximum production was as much as 5 times higher than any other (vertical or directional) well in the field.

SAND PRODUCTION AND CONING

Most wells produce at a high flow rate with a resulting high pressure drawdown. Horizontal wells have a larger section of the wellbore exposed to the formation. Therefore pressure drawdown is less for a given production rate in horizontal wells than in vertical and directional wells. This lessens production problems related to pressure drawdown.

At higher drawdown pressures, sand production is a common problem, especially the production of unconsolidated and fine-grained sand. Sand erodes and plugs equipment and restricts the flow rate. Screens and gravel packing limit sand entry into the wellbore and in some cases reduce production rates. Less pressure drawdown eliminates the need for screens and gravel packing and allows higher production rates.

Water coning problems can be reduced with less pressure drawdown. Water frequently underlies oil or gas in the reservoir. Wells completed in the oil and gas section may produce water by coning. High drawdown causes the water to flow upward, coning into the productive section and thus being produced with the oil and gas (see Fig. 1–9). Water production often restricts the production of oil and gas. Produced water must be disposed of by approved methods, further increasing production cost.

Gas coning occurs in completions in which an oil zone has an overlying cap of natural gas. High drawdowns cause the gas to flow downward, coning into the oil section and thus being produced with the oil (see Fig. 1–9). It is preferable to leave the gas in place to conserve reservoir energy.

Horizontal wells allow higher production rates at correspondingly lower drawdown pressure as described. This reduces the problem of water and gas coning. It is possible to restrict coning further by placing the horizontal lateral in the reservoir in the optimum position relative to the water, oil, and gas contacts.

OTHER APPLICATIONS

Horizontal drilling is highly applicable to existing cased vertical and directional wells with larger diameter casing and under favorable conditions. These wells are already drilled and cased, and reentering them will be a major application of horizontal drilling

Figure 1–9
Oil, gas, and water coning

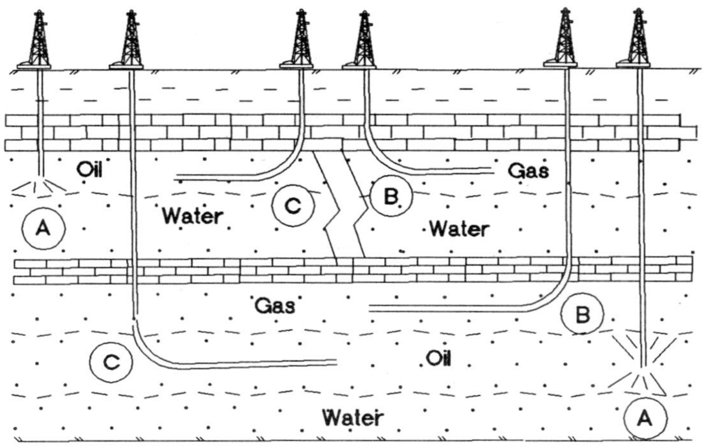

A = Vertical well with coning B = Gas well, no coning
C = Oil well, no coning

due to the large number of existing wells and the lower general costs involved. Many of these wells are depleted, but the higher production from horizontal completions may justify reentry. For example, an abandoned producing well in the North Sea was reentered, drilled horizontally, and completed, doubling production from the field.

New horizontal wells have been successful, so reentering and drilling existing wells horizontally is expected to give similar results. One such potential field is the Pearsall field in south central Texas, which has about 2,000 vertical wells and limited field development because of low productivity. A few of the many other prospective areas include the Niobrara in the Denver Julesburg basin in Wyoming and Colorado, the Cretaceous Mesaverde in Utah and Western Colorado, the Baken shale in the Williston basin in Montana, and the Sprayberry in West Texas.

Horizontal drilling has the potential in secondary, tertiary, and enhanced-recovery procedures to recover part of the remaining oil. Large sections exposed to the formation will increase gravity drainage efficiency. Horizontal drilling should increase injectivity, improve sweep efficiencies, and reduce the number of wells needed for waterflooding and steam injection for recovering heavy oils. It is especially applicable for improving flooding sweep efficiencies, which allows production of oil from isolated areas that were bypassed by flooding from vertical wells. There are very large reserves of heavy oil in the world. This process should be equally applicable in miscible, carbon dioxide, and inert gas floods and some repressurization projects.

A modified form of horizontal drilling places pipelines underneath areas where conventional methods cannot be used. These locations include roads, rivers, ship channels, and industrial areas (see Fig. 1–10).

Horizontal wells should be efficient at producing methane gas from shallow coal beds in the western United States. This also would serve a secondary purpose of reducing the mining cost of drafting to dilute the gaseous mixture in the mine to a safe working level. Other industries benefit from horizontal drilling techniques in different forms, such as the mining industry's use of blast holes.

Combined directional and horizontal drilling may have other applications. These include reduced well spacing, in situ oil shale retorting, coal gasification, in situ leaching in the mineral industry, and heating heavy oil and tar sands. The same general procedures discussed here (and/or modified forms of drilling) apply.

DESIGN GUIDELINES

It is best to design directional and horizontal drilling programs by preparing the optimal well path following the objectives of the program. Guidelines include various controls or limiting param-

Figure 1–10
Pipeline river crossing

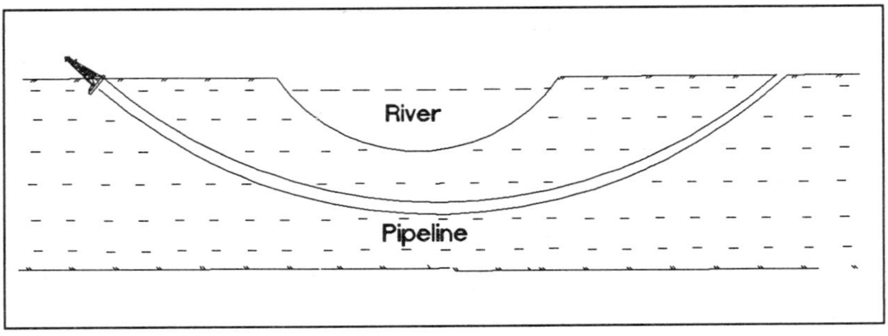

eters based on equipment specifications and experience. Sometimes guidelines require modification because of hole and program requirements. Normally this suggests a higher level of risk. It is best to reduce the risk as much as possible by making the best choice of available factors to reduce risk.

DEFINITIONS

Various terms are summarized here for preliminary clarification and are covered in more detail in the later text. The terms oil and gas are interchangeable for most purposes and drilling operations for either oil or gas are similar. The words well and hole often are interchangeable. Hole generally refers specifically to the drilled hole or wellbore. Well refers to the hole or well after completion. Well is also a collective term referring to the entire rig, wellbore, and drilling site. The terms deviated and sidetracked often are used interchangeably, and the operations are similar (for different reasons) as described in Chapter 3.

Well depth measured along the axis of the wellbore is the measured depth (MD), equivalent to drilled depth. This is used for drilling measurements, casing footage, and other measurements of length along the wellbore. True vertical depth (TVD) is the vertical distance between a point in the wellbore and the plane of the surface (immediately above the point). Measured depth is always equal to or greater than true vertical depth (see Fig. 1–11).

Drift or inclination is the angle between the line of the wellbore and a vertical line, with both lying in a vertical plane. The apex of the angle points upward, and the drift is the angle below the intersection of the wellbore and the vertical line.

Direction or course is the compass or azimuth direction of the horizontal component of a line along the axis of the wellbore. Tool face is the horizontal component of the direction toward which the bit, other drill tool, or whipstock points. Bends are changes of angle in the vertical plane, and turns are changes of angle in the horizontal plane.

This text refers to holes that are either vertical, straight, curved, or a combination of these. Drillholes are seldom exactly vertical, perfectly straight, or precisely curved. Variances of a few degrees are common, the amount depending upon requirements of the specific drilling project, the manner of drilling, and related factors.

GENERAL TERMS

The terms low- and high-angle refer to the drift angle. They are not standardized in industry practice, and general usage is somewhat vague. There is a natural division at a drift angle of about 60°.

Figure 1–11
Depths, angles, and departures

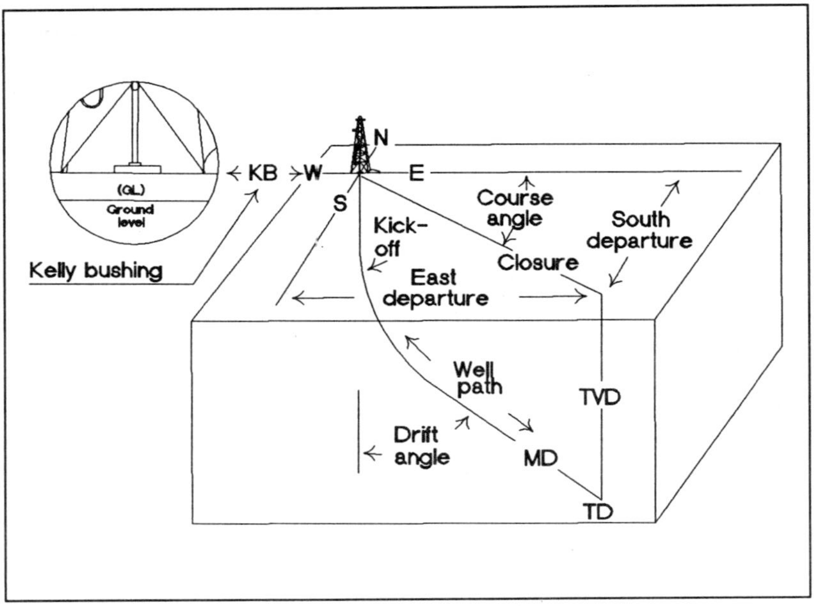

Drilling and operational techniques and problems differ significantly above and below this angle. Therefore, low angles are 60° or less and placed in the directional classification; higher angles are included with horizontal classifications. A similar definition problem occurs in separating extended-reach and horizontal wells. Some operators contend that the drilling degree of difficulty is about the same after inclinations of 70°–80°. Others have arbitrarily separated high-angle directional and horizontal wells at 75° of inclination. Most accept 80° as equivalent to a horizontal well.

Reference information can be very helpful. It is always important to obtain operational information and data from other wells in the area, as well as to review well histories for reference design and operational data. These include problems in building, holding, and dropping angle; performance of various assemblies; and drilling and formation problems. Other sources of information include equipment suppliers, trade journals, and published literature. The importance of researching records and detailed planning cannot be overemphasized.

It is important to simplify the design as much as possible. Directional and horizontal drilling equipment and procedures are well established, but operations are not routine. They take longer to drill than vertical wells. Reasons for this include related and necessary operations such as deviating, making correction runs, circulating, taking surveys, and extra tripping. Also, penetration rates may be slower. These operations frequently take longer than planned. Extended operating time increases risk, and vertical drilling problems increase in directional and horizontal drilling. Problems directly related to directional and horizontal drilling also occur.

Hydraulics must be calculated to ensure adequate mud pressure and volume to operate the turbine or motor and remove drill cuttings. Hole cleaning is a common problem in high-angle and horizontal holes, so it is important to have adequate mud pressure and volume. Calculations should include hydrostatic pressure of the mud column and other pressures based on true vertical depth for high-angle holes. Measured depth commonly is sufficiently accurate in vertical and very low-angle directional wells. There may be appreciable differences between true vertical and measured depth in directional wells, especially with higher angles.

Excess drag and torque can be a major problem (see Chapters 4 and 5). Many directional and horizontal operations such as bends and turns cause increased drag and torque, but they are necessary. It is useful to deviate as deep as possible to minimize the amount of directional hole causing torque and drag problems, and to design for minimum changes of angle and smoothly curved sections. Vertical and straight, inclined sections should be drilled straight, while providing for casing through sections that will cause the most drag and torque. Drillstrings should be designed with adequate overpull, and the design must provide for casing wear. (Formulas for calculating torque and drag are available and may be helpful.) Drilling and tripping cause accelerated wear, especially in bends and turns, so consideration should be given to using heavier weights and higher grades of casing. Normal casing inspection procedures should be followed, and additional inspections may be required in more complex patterns, especially when casing loads are critical.

Regular rotary assemblies limit angle build to about 4°/100 ft and angle drop to about 3°/100 ft. Aggressive assemblies obtain higher rates. Rotary assemblies are most efficient at angles between 25° and 45°. It is crucial not to design for rotary drilling of straight, inclined hole sections with drift angles less than 15°, except for very short sections, because of the difficulty of angle control. The design should use the minimum change of angle, usually in the order of 2.5°/100 ft.

Absolute dogleg is the absolute change of angle in the combined vertical and horizontal directions measured in deg/100 ft. It should be limited to about 4°/100 ft when possible. Higher changes increase the risk of keyseats and other hole problems. Lower build rates allow tools such as packed-hole assemblies to pass without reaming. Reaming should be eliminated whenever possible; it is a high-risk operation, requiring additional time and increasing costs. Extended-reach and horizontal holes often change angle at higher rates with a correspondingly higher risk.

Hole diameters may be determined by the pattern type and, to a lesser extent, operator preference. Optimum hole size is 8 3/4 in. to 9 7/8 in. Acceptable sizes range from 6 3/4 in. to 12 1/4 in. Small holes require smaller motors that are less reliable and efficient. It is more difficult to deviate and drill larger holes, especially in very hard, abrasive formations. It is important to design so that most drilling is in optimally sized holes.

Borehole stability may be a problem in the horizontal hole section, although it is not reported as a major problem in the literature. Special tests and calculations aid in determining this. Sometimes heavier mud is used during drilling, and heavier weight casing later. In practice, some rock movement may be permissible with good designs.

Pilot holes should be designed according to target formation depths and other information. This may save drilling a more costly horizontal hole. Final course adjustments should be provided for with tangent sections as described in the section on tangents later in this chapter. If there is any question about its exact position, the surface location should be surveyed again. Some reasons for this might be a survey of questionable accuracy or inadvertent movement of the location stake while either building the location or moving the rig onto the location.

Casing run in deviated holes is subject to bending and buckling stresses similar to that described for drillpipe in Chapter 5. These can cause a failure under severe conditions. There is less risk of failure in the casing collars because they are stronger than the pipe body. Still, the threaded section on collared casing may be a point of weakness. Casing failures in these instances are uncommon but should be considered when designing the program.

Casing sizes are generally the same as in vertical holes. Common sizes include 7 in., 7 5/8 in., 9 5/8 in., 10 3/4 in., and 13 3/8 in. Intermediate and special sizes may be used for additional casings. The design engineer should consider placing a heavier casing in the deviated section of high-angle holes for additional wear protection, as well as placing additional centralizers through the deviated sections as needed for good centralization during cementing.

The design should allow for an extra string of casing for higher risk wells drilled in hazardous areas, particularly in earlier wells where less information is available or known. Drilling problems are difficult to predict, especially for horizontal and high-angle, extended-reach wells. The casing may be omitted if it is not needed. This procedure can save completing the well at a lesser depth before testing all objective horizons or trying to drill at greater depths in a smaller diameter hole with the resulting problems and higher risks.

Formation evaluation is an important part of planning and designing a well program. The formations should be evaluated on directional wells in the same manner as vertical wells, with allowances made for drift angles. It is important to plan and design carefully for evaluation in high-angle and horizontal holes where more problems occur. Evaluation procedures differ as explained in Chapter 5. The logging features of measurement-while-drilling are gaining acceptance. Coring should be limited because of reduced directional control. Open hole formation testing also should be limited because of the high risk of sticking. Mud logging is common; on most wells it is used to help in drilling, to support hole guidance, and to help in evaluating formations.

Completions should be planned and designed to optimize production rates. This includes considering the type of formation, reservoir pressure, drive mechanism, reserves, stimulation, production lift, long-term economics, and future remedial work.

RISK AND DEGREE OF DIFFICULTY

Drilling operations have two basic classes of risk. One is the risk encountered during drilling and completing the well. The second is the risk that oil and gas may not occur or volumes and flow rates will be less than originally estimated. Both are equally important. They depend upon preliminary investigation, careful planning, and prudent operations. The well must be located where oil and gas occur in economic quantities. Otherwise, the drilling operation is a wasted cost despite operating efficiency.

Risks include excess drag and torque, the possibility of sticking or keyseating, problems within the formations or with the casing, blowouts, and other drilling problems as described in Chapters 4 and 5. Additional risks in directional and horizontal wells relate to the number and radius of bends and turns, inclination, length of the inclined and horizontal hole section(s), wellbore stability, and operator experience. As with many new procedures, mistakes have been made in horizontal drilling, sometimes compounded by the rapid increase in its use and the lack of experience. Improved equipment and techniques and additional experience will reduce risks and associated problems.

A blowout in Texas occurred when a well, while being drilled horizontally, caught fire and destroyed the rig. Most, if not all, similar situations can be prevented with good safety equipment and operating procedures. The severity and likelihood of problems increase with depth and higher angles. Risk is least for vertical patterns, increases with directional patterns, and is highest for horizontal drilling.

The risk of successfully drilling and completing the well relates to the "Degree of Difficulty." Higher risks are associated with a higher degree of difficulty and result in higher costs. Table 1–1 compares the degree of difficulty of drilling directional and horizontal wells, referenced to vertical wells.

Table 1-1
Directional/Horizontal "Degree Of Difficulty."

Pattern Classification	Degree of Difficulty	Relative Cost (% greater than vertical)
VERTICAL (reference)	Low	0.0
DIRECTIONAL		
Single-bend	Low	+ 25
Double-bend	Low to Medium	+ 50
Complex	Medium	+ 100
Extended-reach	Medium to High	+ 150
High-angle	High	+ 200
Slant	Low to Medium	+ 50
HORIZONTAL		
Short Radius	High	+ 200
Medium Radius	Medium to High	+ 150
Long Radius	High	+ 200

The reference well is a vertical hole located in the same area as the directional and horizontal wells. These are approximate and are listed only to give an order of magnitude of risk. THESE SHOULD NOT BE USED FOR ACTUAL ESTIMATES.

WELL COSTS AND ECONOMICS

Well cost and economics depend upon the specific project. Approximate costs of directional and horizontal wells relate to the degree of difficulty as listed in Table 1–1. These are only a rule of thumb covering a broad range. Actual costs depend upon the specific project, pattern complexity, and various problems described in the section about risk. Experienced personnel can estimate reliably, but accuracy may decrease in higher risk operations. The operator should always consider drilling a vertical hole before drilling horizontally because of the higher costs (see Fig. 1–12). Operators experienced in horizontal drilling have cost reductions of 20% to 50% after drilling a few wells in an area, so experience in the area is important.

Economics should be based on drilling and completion costs and well productivity in the conventional manner. Special precautions should be taken when estimating productivity. Unquestionably, there have been some horizontal wells with high productivities. However, sometimes there can be a very high decline rate, so that the well is not economical in spite of its high initial rate. Any production reports used for estimates should be verified. Careful extrapolations of initial production for cumulative recovery calcu-

Figure 1–12
Drilling rate comparisons

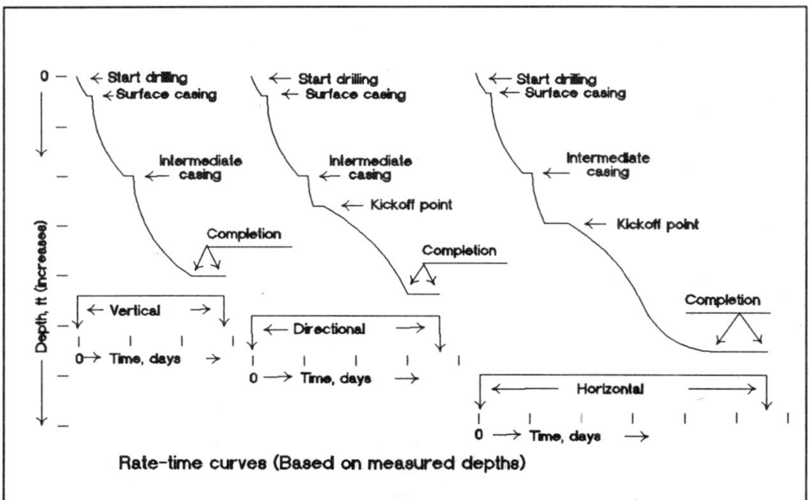

Rate-time curves (Based on measured depths)

lations should be made. It is important to evaluate these correctly, especially before drilling subsequent wells.

DESIGNING/CALCULATING WELL PATTERNS

Well patterns are the various types and combinations of directional and horizontal wells. Common directional patterns are single-bend, double-bend, extended-reach, and slant hole. Complex patterns are the base pattern with one or more bends and turns and various changes of angle (see Fig. 1–13). Horizontal patterns are short, medium, and long turn radius. The turn radius is the radius of the 90° curve (or turn) that changes the direction of the wellbore from vertical to horizontal. These patterns are the most common and considered here as standards. There are other, different horizontal patterns, primarily with different rates of curvature. Combination patterns merge directional and horizontal designs. Common combinations include adding a horizontal section at the end (bottom) of extended-reach and slant hole patterns.

Well patterns are illustrated on vertical and horizontal cross sections as a schematic representation of the wellbore. Complicated designs may use multiple sections for clarification. The schematic illustrates the well path, an imaginary line along the

Figure 1–13
Directional with horizontal and complex patterns

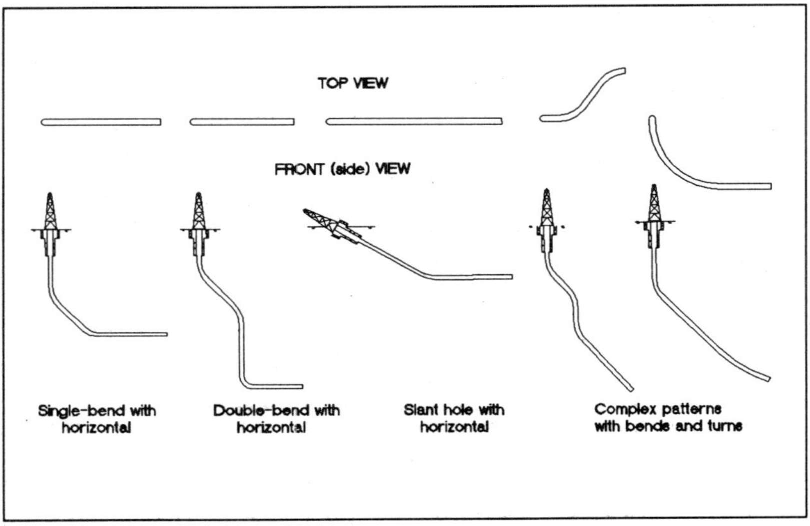

axes of the wellbore. It includes the kickoff depth, the course and angle of the well path, the target and limits, boundary lines, and other relevant features. Normally all calculations are done with computers and schematics are printed or plotted. The well pattern must be designed carefully, paying attention to correct distances and angles.

CLASSIFICATIONS

Three basic well classifications are vertical, directional, and horizontal. Well classifications depend upon the shape of the wellbore, the purpose for drilling the well, and the drilling procedure. Each well classification is subdivided into one or more types or patterns, each serving a specific purpose. Well patterns also identify the different types of wells under the three classifications. Often the name of the pattern is the same as the name of the well type. Vertical wells have a vertical wellbore drilled with standard drilling tools. They represent a majority of wells drilled and are not covered specifically in this text.

Directional and horizontal wellbores are drilled along a planned path through the earth that cannot be drilled by vertical procedures. They are drilled progressively deeper, in any reasonable direction, using special tools and techniques for changing the direction of the wellbore one or more times. Horizontal holes start vertically, curve through a 90° turn, and then continue in the horizontal direction.

Directional and horizontal wells mostly serve separate purposes. Like vertical wells, directional wells locate and produce oil and gas. Horizontal wells produce oil and gas at higher rates and increase total recovery as compared to vertical and directional wells. They also produce economical volumes of oil and gas from some formations that cannot be produced commercially by other drilling methods. The reasons are very significant and explain the acceptance and rapid advance of horizontal drilling.

DIMENSIONAL REFERENCES

Dimensional references are the means of using various measurements of distances and angles to illustrate the well pattern. They locate and define the position of any part of the well including reference depths, well paths, targets, limits, boundaries, and other relevant information. The same depth and point reference system is used in both design and subsequent drilling operations. During the design process, the well plan is plotted as a two-dimensional schematic on the well plat (see Fig. 1–14). Horizontal and vertical cross-sectional views are displayed at convenient scales.

Figure 1–14
Directional well plan

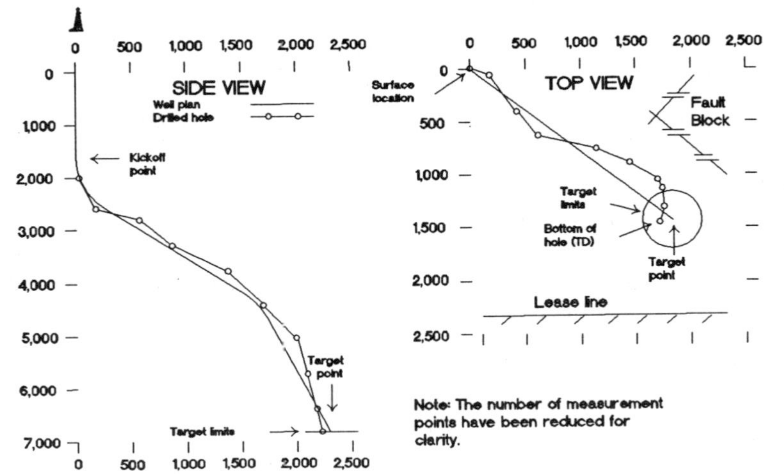

The surface location and elevation must be located precisely by conventional surveying techniques, and ground level elevation is referenced to mean sea level. This is the base reference point for locating all other points in the wellbore. The top of the kelly drive bushing (KB), most often 1 ft above the level of the rotary, is normally the reference point for all depth measurements. It frequently is necessary to convert depth measurements in the hole to sea level reference measurements. The kelly bushing elevation (KBE) is deducted from the depth measurement to obtain the measurement relative to sea level, i.e., above sea level or sub sea level. The measurement of the kelly drive bushing height above ground level (usually 10–45 ft) is recorded for future reference after the rig moves. The top of the surface or first permanent casing head frequently is set at ground level, or its elevation recorded as a permanent future depth reference.

The location of all points in the well are identified by depth and horizontal position referenced to the KB or base reference point unless specified otherwise. Depths are determined as measured depth (MD) and true vertical depth (TVD) as previously defined.

Vertical section is the vertical distance in feet between two points, usually two consecutively surveys.

The horizontal position of a point is measured as rectangular coordinates or departures referenced in horizontal distances from the KB. Coordinates are the shortest straight-line distances from the measured point to the nearest of either the north-south or east-west lines passing through the KB. These are referenced to true north, not magnetic north. For example, the horizontal identification of a point in the wellbore may be "350.25 ft N DEP and 480.62 ft E DEP." The horizontal position of the point is 350.25 ft north and 480.62 ft east of the KB. Closure is the nearest straight-line distance from a point to the surface location measured in the horizontal plane, or 594.70 ft in the example.

Closure and the direction of the line of closure also locate the horizontal position of a point. In the example, the point is identified by line and closure as "594.25 ft E 36° and 5′ N." The point is at a distance of 594.25 ft from the KB on a line that extends from the KB at an angle of 36° and 5' north of east. The same point could be identified as "594.25 ft N 53° and 55′ east" with the line first referenced from the north line. Bearing references are less common. These are similar to the closure and line method except that the angle of the line is always measured in degrees clockwise from true north.

The well path is a line along the axis of the wellbore. It represents a series of points connected by lines. All points should be identified by depth and location referenced to the KB as described. Other points similarly identified are the kickoff point, target, areas, and volumes. Well path limits are the maximum allowable difference in distance between the well plan and the actual well path during drilling. Conventionally, a cylindrical shape along the well path defines well path limits. The radius of the cylinder is the maximum variance (see Fig. 1–15).

The target is the drilling objective. A target in thin formations (about 15 ft thick or less) is represented as a point. The target limit is a circle with the target point as the center and a radius equal to the allowable variance. Thicker targets are delineated as lines with cylindrical shape limits similar to the well path limits. Two or more targets are represented individually at their respective depths.

Hard lines identify areas that cannot be drilled. Lease boundaries and nonproductive areas such as fault blocks should be identified as a line on the horizontal section that cannot be crossed by the drill bit. Adjacent wellbores also are identified with limits beyond which drilling should not occur.

Figure 1–15
Well path, target, and limits

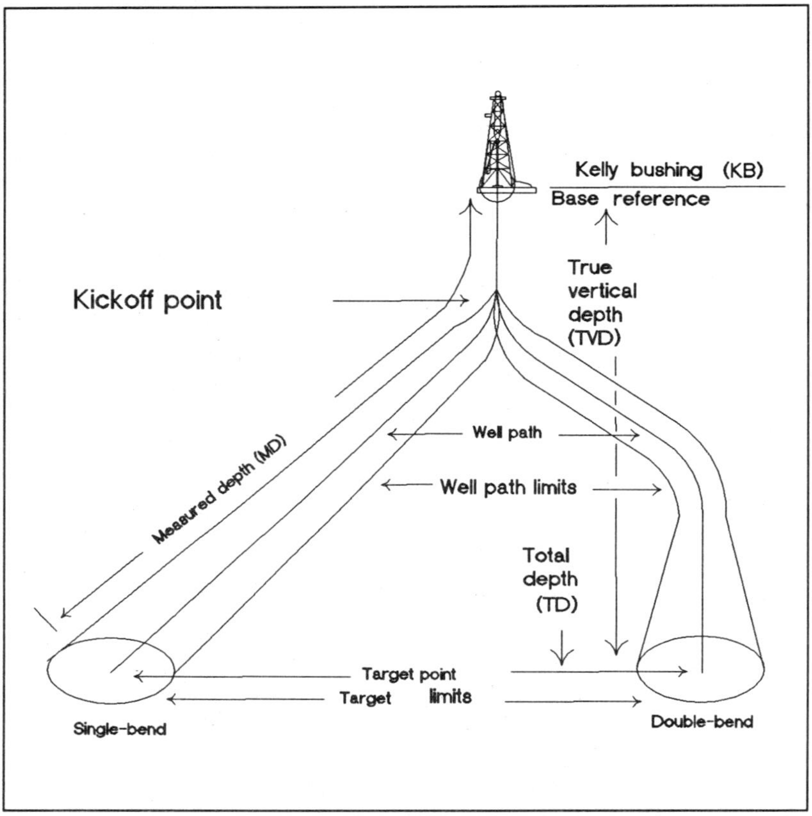

CALCULATIONS

The position of the wellbore at any point may be calculated using formulas and measurements of angles and distances. The data points representing the well path, target, etc. should be set during design, then the reference data calculated. Commonly, computer programs are used to generate the well path and all other reference points and measurements based on guideline input data. The computers also drive printers and plotters that print schematics of the wellbore. The basic procedure still includes calculating the required parameters between two points by one of several formulas listed in Table 1–2.

Table 1–2
Course Calculation Methods.

Average Angle	Minimum Curvature
Balanced Tangential	Radius of Curvature
Callas's Helical Arc	Quadratic
Circular Arc	Tangential
Mercury	

The minimum curvature method is theoretically the most accurate and most commonly used. It is an involved procedure and normally calculated with a computer.

The average angle method is easier to calculate and may be used for preliminary field calculations if a computer is unavailable. It is slightly less accurate by a few percentage points but is acceptable for field work. A hand-held calculator or portable computer at the well site can be used to make calculations that are plotted on a field copy of the directional drilling design. This provides a comparison of actual drilling results with the projected results, so changes can be made immediately as required.

In the general procedure, calculations between two points are made and recorded. The first point is the base reference point or the kelley drive bushing. The position (horizontal location and elevation) of this point is known. Here the reference base point will have a vertical drift and a "zero," or no direction, used in the first calculation. Subsequent points will have both drift and direction as explained in the following. After drilling the well deeper for some distance, a new or final point is selected at some measured depth below the first point. The drift and direction of the hole at this second point is recorded. The drift and direction at both points and the measured distance between them are used to calculate the changes between the two points.

As shown in Figure 1–16, the calculated changes between the two points are vertical section, **CB,** departures, **EC** and **DC,** and closure, **AC.** Each calculation gives the incremental change, either an increase or decrease, from the first point to the second or new point. The changes are added to the known depth and position data from the first point to give the depth and position of the second point. This locates the newer point precisely in relation to the base reference point.

For example, the changes in departures give the coordinates of the last point. This last or new point then becomes the first point

Figure 1–16
Calculating the well path

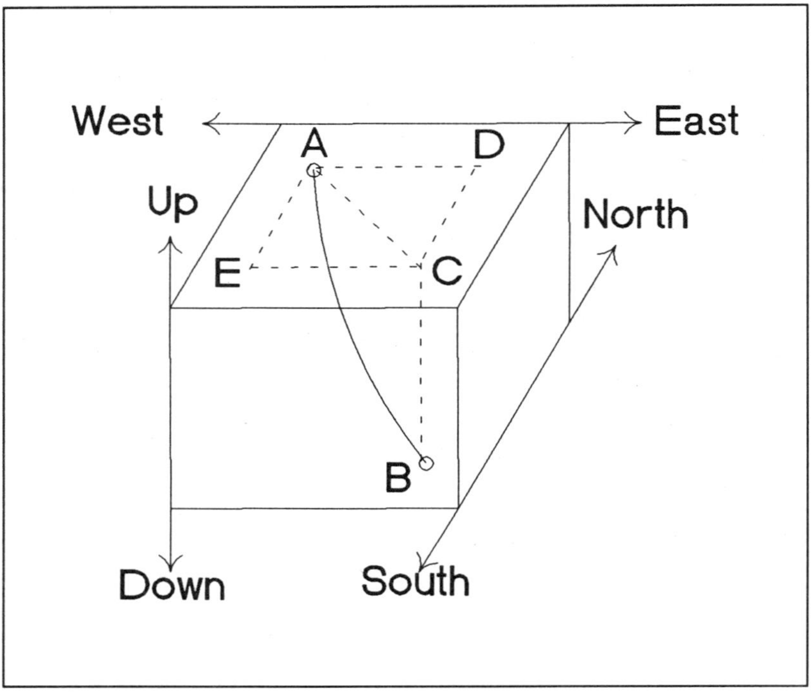

(or temporary base reference point) for calculations after drilling to a new depth at the next "second or new point" and measuring drift and direction. The position of each succeeding point is calculated similarly while drilling the well deeper.

All compass-type magnetic drift surveys or direction measurements reference to magnetic north. Well plats, land title schematics, and other permanent records reference to the geographical true north or true bearing as a universal standard. Therefore, magnetic compass measurements must be corrected from magnetic north to true north so that the well plat will conform with surface and related maps.

The direction and variation in degrees between true and magnetic north depend upon the physical location of the point of measurement, in this case the well site. Magnetic declination charts (isogonic charts) are area maps overlain with lines of equal magnetic declination. The correction is taken from these charts at the measurement location (the well site). The correction is added or subtracted from the magnetic compass reading based on magnetic north to give the corrected direction referenced to true north. Sometimes these corrections are large, ranging from 0 to greater than 20% variance (±) from true north over the continental United States. Magnetic declination changes constantly. The change is very small, but updated values must be used. Many companies have magnetic declination values stored in their computers with programs for correcting magnetic measurements. Note that gyroscopic measurements may be referenced to true north, making correction unnecessary.

Offshore wells in federal waters (outside of state waters) should be corrected to Grid North. Localized areas are defined within a grid system that has specific latitude and longitude selected as the corresponding X and Y axes. A grid correction is applied in order to correct magnetic directions. Wells in international waters mostly use the Universal Transverse Mercator (UTM) grid zone system, which covers broad areas referenced to meridian lines.

KICKOFF POINT

The kickoff point (KOP) is the depth or point in the hole where deviating or sidetracking begins. Kickoff points should be selected to provide an economical, drillable well path into the target. Standard criteria are used and modified subject to the well pattern and any special requirements due to the drill site location.

The KOP should be selected as deep as reasonably possible. Vertical holes can be drilled faster and more economically with fewer problems compared to directional holes. The deeper KOP also may allow vertical clearance to sidetrack higher in the event the first deviated hole section is lost. Deviating at greater depths saves drilled hole. Deeper kickoff points can alleviate other problems such as difficulties with hole cleaning and running logging tools, and casing and production problems after completing the well.

However, there are exceptions. It may be necessary to kickoff at shallower depths if the deeper kickoff point requires higher than normal angles and if the section will be covered later by intermediate casing. Kickoff at shallow depths can be accomplished by jetting or nudging (see Chapter 5) if the formations are very soft and there is sufficient distance to the target.

The KOP should be at least 100 ft below the bottom of the last casing in the hole and preferably 200 ft or more, especially below surface or shallow intermediate casing. This reduces the risk of excess casing wear or splitting the casing shoe. The setting depth of the casing may be adjusted if necessary when the KOP is critical;

the casing may be set higher, or the casing may be set in the hole after deviating.

The KOP should be at least 50 ft, preferably 150 ft, above the top of a fish. Otherwise, the deviated hole may be drilled back into the fish or may reenter the original hole. Either will require a second plugback and sidetrack. The KOP may be located closer to the fish in critical situations by using an assembly with a high-angle build rate. This also increases the risk of a dogleg or crooked hole.

It is easier to deviate or sidetrack in some formations than in others. Gathering information about the formations is one good reason to review all available data on other wells in the area. Very soft formations may increase the difficulty of deviating and building angle. The deviating tool must exert a side force on the formation to cause the hole to deviate. Very soft formations may not have sufficient strength to exert the required counterforce. Therefore, the fulcrum (or back side) of the directional assembly will push into and may partially enter the wall of the hole, providing insufficient lateral thrust. This reduces efficiency, making it more difficult to deviate or sidetrack, build angle at a satisfactory rate, and otherwise control the direction of the deviated hole in softer formations.

Very hard formations, especially abrasive formations, are difficult to drill. Deviation assemblies are less rugged, so bit weight is reduced. This restricts operations, increasing the time spent deviating. It is important to avoid very soft, very hard, abrasive, or laminated formations. The KOP should be selected in medium-soft or medium drillability, massive formations when possible.

The horizontal position of the KOP must be known with reasonable accuracy. Normally, new holes have drift and direction measurements for calculating the KOP. Old holes with casing may not have been surveyed, or surveyed only with a drift instrument. A gyroscopic wellbore survey should be run to determine a precise location. Sometimes a precise location may not be necessary, particularly with large targets. A "cone of uncertainty" often is acceptable in these cases. The horizontal displacement should be calculated for each drift survey. These should be totaled, ignoring direction. The sum is equal to the radius of the cone of uncertainty. It is the maximum possible displacement of the KOP from the surface location, assuming accurate, representative, original measurements.

The exact displacement is unknown but is probably considerably less because of the spiraling tendency during vertical drilling. For example, assume the circle of uncertainty is 60 ft in diameter and the allowable diameter of the target is 600 ft. In this situation, the direction of the deviated hole is controlled for drilling into a target that is 480 ft in diameter, a reasonable size of target in many patterns. The diameter of the new target is equal to the allowable target diameter, reduced by twice the radius of the cone of uncertainty. This can save the time and cost of running the wellbore survey if the variance is acceptable. The procedure is especially applicable to large targets and less difficult patterns. It also is acceptable to some state regulatory agencies.

TARGET

The target is the drilling objective. The size of the target is very important from the viewpoint of cost. Directional drilling technology has advanced to the point where a hole can be drilled into a target a few feet in diameter. Drilling into the casing of a blowout well with a kill well is an example. However, small targets can increase significantly both drilling time and total costs, so the maximum permissible target size is selected. A standard acceptable directional target is a circle 250 ft in diameter at 5,000 ft, 500 ft diameter at 10,000 ft, etc. The maximum permissible target size is always used.

Targets may have elliptical or oblong shapes. When possible, a surface location or program design should be selected so that the long dimension of the target is perpendicular to a horizontal line between the surface location and the target. This may reduce correction runs with rotary assemblies, because it is easier to control the angle than the direction. Directional wells on land often have some flexibility in selecting the surface location. This should be considered in order to improve the pattern. Geological information obtained during drilling may permit increasing the target size, or it may require decreasing the size of the target or moving it in a more favorable direction.

Targets for relief or kill wells range from a few feet to more than a 50 ft radius for an open hole condition. It even may be necessary to penetrate the casing of a cased hole. A less common target is a cylinder, usually oriented vertically. The standard cylindrical target preferably should have the same horizontal size as the recommended directional target. Horizontal hole targets are mostly vertical, normally entered by drilling horizontally into a formation. Vertical control is critical, but there often is more latitude in the horizontal direction.

Single targets are more common for directional and horizontal wells. It is possible for some directional wells to have multiple targets, but there are seldom more than two. These can be at different depths and horizontal positions. Guidelines for single targets apply to multiple targets as well.

DIRECTIONAL DESIGNS

Directional well classifications are subdivided into standard patterns including single-bend, double-bend, extended-reach and slant hole (see Fig. 1–17). Complex patterns have multiple bends and turns. Each well pattern is for a specific purpose, so pattern selection depends upon the reason for drilling the well. The well path should be designed by calculating the changes of angle and length of the straight, inclined section required to connect the kickoff point to the target.

The process starts by selecting the minimum angle of build or drop required to drill the hole into the target. Designs include both deviating and sidetracking, as described in Chapter 3. Holes with these patterns are drilled in various sizes to measured depths of greater than 18,000 ft (shallower for more complex designs). If there is a choice, the design for the most economical type of assembly should be chosen. The difficulty of drilling directional wells increases with increasing angle and depth. Complex patterns with higher angle build and drop rates and more turns and bends are harder to drill. Directional and horizontal patterns can be combined for some drilling situations.

SINGLE-BEND

Single-bend patterns have a single bend in the vertical plane, sometimes called bend-and-run. The pattern starts with a vertical

Figure 1–17
Directional patterns

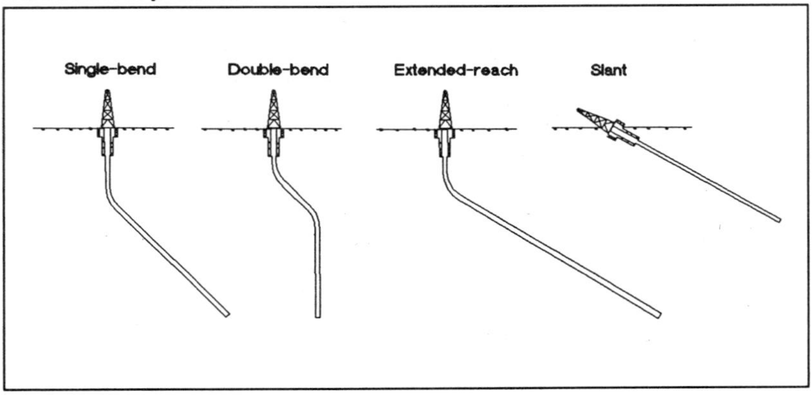

hole. The next step is to deviate or sidetrack at the kickoff point and drill a smooth, upward curve at an increasing angle. Normal angular build rates are 1.5°–2.5°/100 ft, with higher build rates in holes with higher angles. The curved section should be drilled to an inclination normally between 25° and 60°. This drift angle is maintained while drilling a straight, inclined hole into the target. The angle buildup and the drift angle of the straight, inclined sections depend upon the vertical and horizontal distances between the kickoff point and the target. Drilling this pattern is somewhat troublefree and is classified as a low level of drilling difficulty.

This pattern is commonly used to drill multiple wells from a single surface location by placing the conductors close together. It is also used for sidetracking and changing the bottomhole position, for reasons including: bypassing a deeper fish; moving the bottom of the hole updip to avoid water or downdip to avoid a gas cap; by crossing faults; penetrating attic oil or gas or basement oil; and other similar situations. Relief (kill) wells are drilled also to control blowouts. This pattern is used also to drill vertically through problem formations, followed by deviating with a higher angle at a deeper depth. The pattern also serves as a basis for extended-reach and horizontal well patterns.

DOUBLE-BEND

Double-bend *(S)* patterns have two bends in a vertical plane separated by a straight, inclined section. First it is necessary to deviate from a vertical hole, and then drill the angle buildup and the straight, inclined sections similarly to the single-bend pattern. The next step is to drop angle and drill a smooth curve in the downward direction. The angle should be dropped at rates of 1.5°–2.5°/100 ft, and then dropped to vertical. This is followed by drilling vertically downward into the target for standard patterns. It is best to design for drilling with rotary assemblies when possible, especially for the downward curving section.

A common variation has another change of angle in the lower section for drilling a second straight, inclined hole section into the target. Changing the angle in the horizontal direction is also common.

Angle-build and angle-drop rates and the drift and length of the straight, inclined sections should be designed based upon the horizontal and vertical distances between the kickoff point and target(s). High torque and drag may limit depth in complex patterns with multiple bends and turns. This pattern has a moderate to high level of drilling difficulty depending upon the number of bends and turns.

The double-bend pattern is used for similar reasons but often in more complex situations, usually related to the distance and relative position of the kickoff point and target. Uses include drilling multiple targets or long vertical targets, sidetracking a shallow fish, bypassing intervening obstacles such as other wellbores and lease limitations, and penetrating updip or downdip reservoirs. The double-bend is a common base pattern for more complex designs.

EXTENDED-REACH

Extended, long-reach, patterns have one bend in the vertical plane similar to the single-bend pattern. The main difference is a longer, straight, inclined section, often at a higher drift angle for drilling into targets located long horizontal distances from the surface location. The difference between single-bend and extended-reach patterns is not well defined. An arbitrary definition of extended-reach is a horizontal separation between the surface and bottomhole location greater than 3,000–4,000 ft.

Extended-reach patterns should be designed similarly to single-bend patterns with allowances made for a longer straight, inclined section and higher angles. Extended-reach wells have been drilled to measured depths of almost 18,000 ft with horizontal, surface-to-target separations of more than 15,000 ft and at high angles (approaching 80°). Torque and drag increase with depth and may limit the total depth of the well, thus the pattern should be designed to alleviate the condition whenever possible. Extended-reach patterns are combined frequently with horizontal patterns, and in these cases the design of the straight, inclined section often is similar to horizontal laterals as described in the section on horizontal wells.

SLANT HOLE

Slant holes start from the surface at an angle of 30°–45° by drilling with a slant-hole rig. The surface or conductor casing is set at shallow depths, and the remainder of the hole is drilled straight, in an inclined direction. Alternately, it can be deviated to change the direction from a few degrees to horizontal, sometimes a few degrees above horizontal. General design of the pattern and casing strings is similar to other directional holes with allowances made for the angles and tubular compression due to the pull-down system.

Slant holes may have high drag, restricting tubulars from falling freely due to gravity. Slant-hole rigs have a pull-down system (pull down) for pushing the drillstring into the hole during tripping when it is needed. The pull down also helps deliver additional weight to the bit for drilling and is useful when running casing. The pull down creates a downward force, so the drill tools and casing may be in various states of compression. This must be provided for when designing the drill tools and casing.

Slant holes penetrate productive zones at shallow depths at relatively long horizontal distances from surface locations. This is similar to a specialized application of extended-reach patterns and serves the same purpose. The shallow depth limits the horizontal distance obtainable with conventional extended-reach patterns. Extended-reach wells require some vertical distance in order to change the vertical direction of the hole. Slant holes start at an angle, so they drill longer horizontal distances into targets at shallow measured depths.

HORIZONTAL DESIGNS

Horizontal designs are well plans with a section or lateral drilled horizontally through the earth. Conventionally, these wells deviate at the kickoff point, drill through a 90° curve and then drill horizontally into the formation. They may be drilled as new wells or in older, cased holes, if the casing diameter is sufficiently large. Horizontal drilling is applicable in a wide range of depths and sand thickness. Measured depths of 10,000 ft are somewhat common, with some at depths greater than 14,000 ft. Horizontal laterals have been drilled more than 2,500 ft into thin sands (less than 10 to 15 ft thick) and nearly 2,000 ft into slightly thicker sands at depths greater than 10,000 ft.

It is also possible to drill horizontally as an extension of a directional pattern, including extended-reach and slant holes. The surface location of the directional well is selected and then drilled so that the bottom of the wellbore is near the desired target point. Then a curved section is drilled until the hole is horizontal, followed by drilling horizontally laterally into the formation. One high-angle extended-reach well had a total horizontal displacement of nearly 13,000 ft, including 5,500 ft of horizontal hole. Another had a total horizontal displacement of more than 16,000 ft, including more than 1,500 ft of horizontal section. There are combination wells in most major fields, and they are common in offshore operations.

The horizontal classification is subdivided into patterns based on the length of the radius (turn) of the 90° curved angle-build section (see Fig. 1–18).

Table 1–3
Horizontal Pattern Classifications.

Pattern Name	Turn Radius, ft	Build Rate ° /100 ft	Horizontal Extension, ft
Short	2–60	1,000+ –95	100–800
Medium	300–800	19.1–7.2	1,500–3,000
Long	1,000–3,000	5.7–1.2	2,000–5,000

Angular build rates are in degrees per 100 ft of measured depth. Horizontal classifications are not standardized in the industry. Table 1–3 contains a summarized average of classifications used by various operators and service companies. These are guidelines within a wide variation of angle-build rates. There are gaps between the pattern ranges in Table 1–3. It is more difficult to drill in the gap areas because of equipment limitations, and it is naturally easier to drill within the pattern ranges. A few wells are drilled outside of the pattern ranges, but most are drilled within the ranges listed in Table 1–3.

The turn radius of about 300 ft is a natural division between short- and medium-turn patterns for several reasons. It is about the minimum turn radius that most standard tubulars can pass through safely with careful handling. Most shorter-turn curves require special articulated or smaller diameter tubulars. Standard deviation tools cannot build angle at higher turn rates in a controlled manner. The ability to use standard tubulars and deviation equipment is important for conducting efficient operations and controlling costs. The difference between medium- and long-turn patterns is less well defined.

Design procedures for all horizontal hole classifications are similar. First it is necessary to evaluate the oil- and gas-bearing strata carefully. The next step is to select the correct length for the horizontal section and find the best position for the horizontal section in the reservoir, including areal location, direction, and depth relative to formation boundaries. The horizontal section is often placed parallel to fluid interfaces and perpendicular to fractures. Sometimes the horizontal section is oriented based on

Figure 1–18
Horizontal patterns

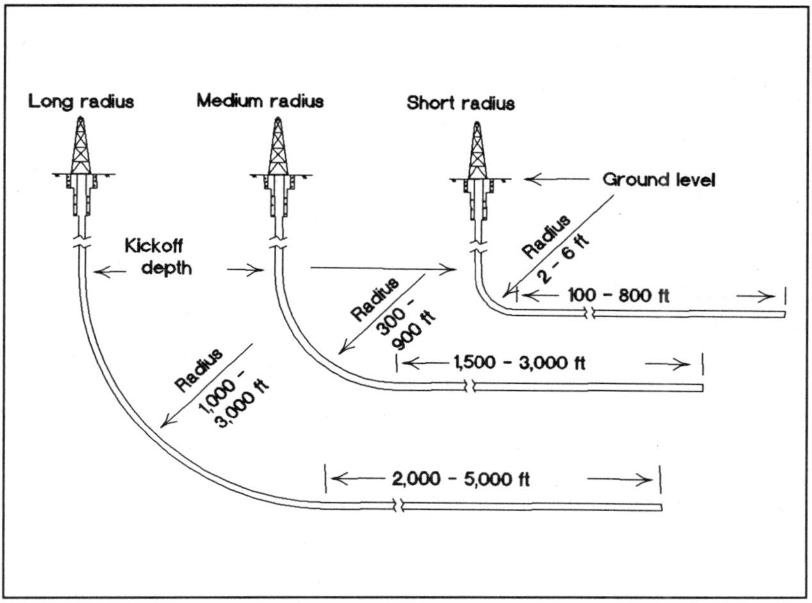

analysis of fracture propagation. This ensures the most efficient fracturing at completion. The reasons for drilling the horizontal well as described earlier in this chapter often determine the position. These factors determine the true vertical depth to the horizontal lateral and its length and position in the reservoir relative to the surface location.

It is important to evaluate the advantages and disadvantages of the range of turn radii and to select the one most applicable to the well under consideration. The kickoff point is equal to the true vertical depth to the horizontal section less the length of the turn radius. The surface location should be positioned a distance equal to the turn radius from the point where the hole becomes horizontal. The hole size of the curved and horizontal sections is chosen for optimum operations. The vertical hole normally is a standard size larger.

The design engineer should also provide for a tangent section (two in areas with less information and possibly with thin reservoirs). The measured depth is calculated, and a cross-sectional diagram is drawn to scale. It is important to verify that drilling assemblies can drill the pattern efficiently. Plans should call for drilling the curved section with a drilling assembly that builds angle at the rate required by the pattern design. Preference should be given to drilling with steerable assemblies as often as possible. The design is completed by selecting the path and target limits, casing points, and completion procedures.

The procedure is modified slightly for drilling a horizontal hole when reentering an old well with casing. The smallest casing size determines the maximum size of the horizontal drill tools. This may limit the turn radius and the resulting length of the lateral hole section. A turn radius should be selected based on the size of tools available and the procedure for deviating through the casing. It is important to provide for plugging back and removing a section of casing (see Chapter 3). Then the design is completed as described.

The applicable turn radius is selected by evaluating various factors. A longer vertical section is easier to drill but requires a shorter turn radius for a given depth to the position of the horizontal section in the formation. It is more difficult to drill a shorter turn radius because of the higher angle-build rate as compared to a larger radius turn (see Fig. 1–19). Problems with hole cleaning and high drag and torque increase with increasing measured depth, such as for a longer turn radius. It is helpful to have a good understanding of the design and use of bottomhole assemblies.

SHORT-TURN

Short-turn patterns, sometimes called drainholes, are drilled in existing, cased wellbores. They have a short turn radius of a few feet to about 60 ft and build angle at very high rates. Several horizontal holes may be drilled from the same wellbore. The average maximum length of the laterals is about 300–700 ft in the optimum case, but generally is considerably shorter.

Short turn radius patterns are less common, partially due to inherent disadvantages. The procedure requires milling a section of casing. Special pipe is required to drill the short turn radius. Horizontal laterals are somewhat short and may be drilled without directional control. The short turn radius and the small hole size limit completion procedures. Special drilling equipment and procedures can be complicated.

The pattern is not applicable in all situations. Successful short-turn projects cost less than those with a larger turn radius but give smaller increases in production. Smaller targets can be penetrated more accurately because of the equipment used and the short turn

Figure 1–19
Buildup angle vs. turn radius and measured length of curved section

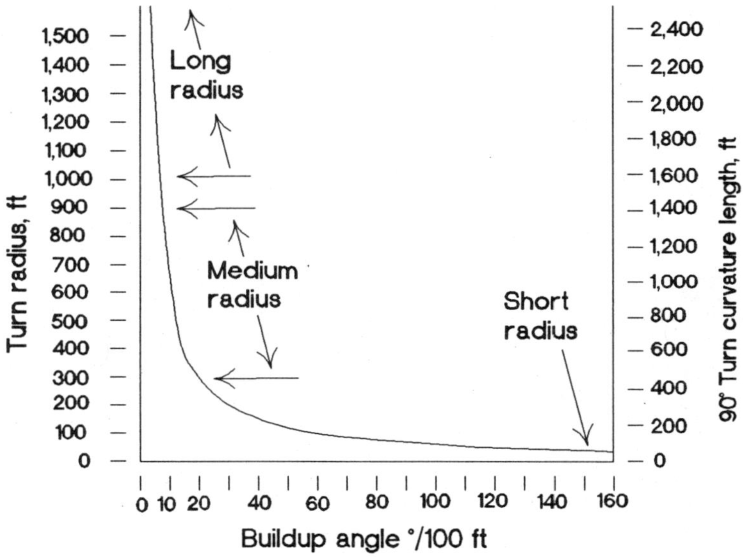

radius. The curve turns in a minimum horizontal distance, so the pattern is applicable in areas such as a small lease with limited space. The short curvature allows placement of artificial lift pumps closer to the reservoir. This increases production efficiency in reservoirs with lower pressure. It also may serve as a pilot program for determining the applicability of drilling horizontally with longer horizontal sections.

The design includes removal of a section of casing by milling. Articulated or small diameter drillpipe can be used to sidetrack off special whipstocks. Most drainhole equipment has a fixed buildup rate, so the vertical depth to the horizontal lateral determines the kickoff point.

MEDIUM-TURN

Medium-turn holes are the most common horizontal drilling pattern, especially on land operations. They have a turn radius of 300–800 ft, corresponding to angle-build rates of 19.1°–7.2° /100 ft MD. Horizontal laterals average about 1,500–3,000 ft in length with maximum penetrations of more than 4,000 ft. The pattern is very flexible and applicable to most drilling conditions encountered, including deeper holes, high pressures, and formation problems. Horizontal sections have been drilled in cased wellbores below 14,600 ft TVD; two horizontal laterals, about 3,000 ft and 2,000 ft long and about 180° apart, were drilled below 7 in. casing from the same wellbore.

Most wells are drilled in open holes with diameters between 7 in. and 9 in. Wells with a longer turn radius in the upper end of the classification may have larger hole diameters of up to about 12 1/4 in. The shorter turn radius is used for sidetracking in cased holes with larger casing, usually with diameters of 7 in. to 7 5/8 in. or larger. Smaller hole sizes are selected for the shorter turn radius and drilled with slim-hole tools and techniques. Drilling with split drilling assemblies reduces torque and drag, and increased bit weight is used in applicable situations. Steerable assemblies are used when possible, and measurement-while-drilling is used most commonly.

Sometimes information about the formation and precise depths is unknown. A vertical hole can be drilled through the target horizon(s) first for logging and evaluating the formations. Then, if justified, the vertical hole can be plugged-back, and the curved and horizontal sections can be sidetracked and drilled. This can save the high cost of drilling the horizontal section if the formations are not productive. This is more commonly used for exploration wells and for wells drilled along the edge of a reservoir. Tangent sections may be used as described in the section on tangents.

LONG-TURN

Long-turn patterns have a turn radius of 1,000–3,000 ft, corresponding to angle-build rates of 5.7°–1.2° /100 ft MD. Horizontal laterals average about 2,000–5,000 feet in length with maximum penetrations of more than 5,700 ft. This pattern is usual in horizontal drilling, especially in offshore operations where long horizontal displacements are common. The pattern is applicable in most drilling conditions, including in deeper holes, under high pressures, and wherever formation problems occur. It seldom is used for reentering older cased holes because of larger hole sizes and the possibility of deviation at shallower depths. The pattern is common offshore for drilling multiple wells from a single drilling platform with long horizontal displacements. Designing these patterns may be more difficult because deeper holes increase exposure to drilling problems. The most serious problems are high drag and torque, and cleaning the hole efficiently.

Average wells in long turn radius patterns are generally deeper than those of other patterns, so larger casing sizes and large diameter holes (up to 12 1/4 in.) are used. This helps minimize problems and improves well control. It allows the use of standard tool sizes in the deeper sections and provides for an extra string of casing if unexpected drilling problems occur. Some deeper patterns require larger hole and casing sizes in the shallower deviated hole section. It may be preferable to drill a 12 1/4 in. deviated hole and open it to the larger size depending upon drilling conditions and depth. Deviating in larger size holes often is difficult, especially in harder formations. Long turn radius holes commonly have longer horizontal displacements. Drilling is done sometimes with rotary assemblies but usually motor assemblies are used, especially in the deeper sections. Conventional tubulars are used, as well as more casing and liners because of greater depths.

Casing programs depend on many factors, including turn radius, formation stability, and length of the curved and horizontal sections. Deeper wells with a larger turn radius and longer laterals often have additional casing or liners. Intermediate casing may be set in the middle or near the end of the curvature for deeper holes. Casing frequently is set near the middle of the curvature for very long-turn holes. This reduces drag and torque while drilling the final buildup section. A short, straight inclined section is drilled below the casing shoe before continuing to build angle. This reduces wear at the casing shoe and minimizes the risk of split casing. Casing may be set after the curved section is drilled, which minimizes problems from the upper hole while drilling the horizontal section.

Formation data and precise depths often are necessary. In some cases, a straight, inclined section is drilled starting near the middle or latter part of the curved section and through the prospective formations. This hole serves the same purpose as straight holes in medium radius patterns. The formations are evaluated, marker beds are identified, and precise depth measurements are obtained. The next step is to plug back, sidetrack, and drill the remainder of the buildup section and the horizontal section. Tangents help to enter the target accurately as described in the section on tangents later in this chapter. Measurements are taken while drilling and sometimes the associated formation logging feature can be an important guide.

HORIZONTAL SECTION

The horizontal section (lateral) is drilled into the reservoir at an angle of about 90°. This important section is in the oil and gas reservoir and is a major factor in determining the success of the well. Long laterals are drilled because oil and gas production normally increases with increasing length (up to certain limits). Risk also increases because of the greater frequency of problems while drilling horizontally. These factors are evaluated and the optimum lateral length is selected. Sometimes computer simulator programs can help to determine the length of the lateral if there is sufficient information known about the formation. If not, drilling a vertical or tangent into the formation to obtain the information may be justified.

The vertical position of the lateral (and sometimes its direction) are important. The lateral should be positioned to maximize either oil and/or gas production. Some reservoirs have fractures oriented in one direction. A lateral placed perpendicular to these will intersect more fractures and have correspondingly higher productivity. The lateral should be placed in permeable areas that have higher flow capacities, subject to the type of production. Some reservoirs have directional permeability that may be a factor, and reservoirs may have lower permeability near one or both boundaries that must be taken into consideration.

The thickness of the formation, the contents of oil, gas, and/or water, and the heights of the fluid columns are major considerations in lateral placement. The lateral should be placed in the middle or upper third of most thin reservoirs (10–20 ft thick), depending upon fluid contents. The lateral should be placed near the top of gas reservoirs with underlying water. The presence of gas condensate also may be a consideration. If so, the lateral should be placed near the bottom of oil reservoirs without water, and higher in the oil column if water is present. Lateral position relative to the oil/water contact depends upon the risk of coning.

The vertical position of the lateral may be more critical because of the risk of coning in reservoirs containing some combination of oil, gas, and/or water. Lower viscosity fluids will preferentially flow (cone) into the wellbore under the same conditions of pressure drawdown. The normal order of low to high viscosity is gas, water, and oil. Therefore, the risk of water coning into the wellbore increases if the lateral in the oil column is near an underlying oil/water interface. On the other hand, a lateral that is too high in the oil column may reduce oil production or ultimate recovery. If the reservoir has a gas cap, then gas can cone downward into the horizontal wellbore if it is too close to the gas/oil interface. Water underlying the gas will cone into the wellbore under conditions of high drawdown when the lateral is close to the gas/water contact. Active water drives and formation dips may affect coning. The position of the lateral relative to the oil/water, gas/oil, and gas/water interfaces must be located based on experience in the area, including the results of other operators and, to some extent, computer simulations.

It is normal to plan for the lateral position when designing the program. At times it may depend upon specific conditions determined during drilling. In summary, the lateral should be positioned in the reservoir after carefully reviewing all related factors and other conditions specific to the wellbore under consideration.

TANGENTS

Tangents are relatively short, straight, inclined sections drilled in the angle buildup section of medium- and long-radius holes. They provide for final course adjustments while drilling the lower section of the hole so that it becomes horizontal at a precise depth. Unexpected items must be allowed for that might cause or require directional changes of the lower hole. These include geological information obtained while drilling the hole, such as formation depth changes, variable thicknesses, and areas where assembly build or drop rates are difficult to predict or control.

Tangents are especially important when drilling horizontally through thin formations and in other cases where lateral placement is critical. Tangents provide a means to correct for these items. They also allow flexibility to drill upperhole sections more quickly with less attention paid to build rates, since later corrections are possible with the tangent.

Initial designs should be for fixed length and angle tangents, since these can be adjusted as needed during drilling. Tangents should be planned for, even after drilling a vertical section for target formation depth or other information. The horizontal hole may enter the formation a considerable horizontal distance from this point. Tangents may be used under favorable conditions to place horizontal holes through two formations separated by vertical distances of more than 100 ft in the same wellbore. The tangent procedure is very effective. It allows entering formations less than 10 ft thick at a true vertical depth of about 10,000 feet.

Tangents may require several extra trips for changing bottomhole assemblies and time for drilling the tangent. An upper tangent may be located in long turn radius holes at the point where the curve has built to about 45°. The lower tangent can be located near the end of the buildup section and serves the same purpose as the upper tangent. It is more common, especially drilling into a thin vertical target. It is possible to omit tangents sometimes when drilling with a steerable assembly. A possible compromise is to omit the upper tangent and drill a single tangent with a relaxed build rate lower in the curved hole section. Tangents are used commonly in the first few wells in an area and omitted in subsequent wells because of known depths and drilling conditions. This expedites operations and reduces costs.

BIBLIOGRAPHY

N. Adams and L. G. Kuhlman. "How to Prevent or Minimize Shallow Gas Blowouts." Part 2 *World Oil* (June 1991): 66ff.

P. J. Becker. "Texas Eastern Tests Directional Drilling in Little Missouri Crossing." *Oil & Gas Journal* (April 25, 1988): 48–91.

W. B. Bradley. "Factors Affecting the Control of Borehole Angle in Straight and Directional Wells." *Journal of Petroleum Technology* (June 1975): 679.

G. M. Briggs. "How to Design a Medium-Radius Horizontal Well." *Petroleum Engineer International* (September 1989): 26–37.

L. Bruckert. "Horizontal Well Improves Oil Recovery from Polymer Flood." *Oil & Gas Journal* (December 18, 1989): 35–39.

R. M. Butler. "The Potential for Horizontal Wells for Petroleum Production." *Journal of Petroleum Technology* (March 1989): 39–47.

N. P. Callas. "Computing Directional Surveys with a Helical Method." *Journal of Petroleum Technology* 24 (August 1972): 935–943.

M. Chambers and J. Hanson. "Australian Well Breaks Horizontal Displacement Record." *Journal Of Petroleum Technology* (February 1984): 241–248.

D. B. Christian. "Planning and Operational Requirements for a Shallow-Objective, High-Angle Well in the Gulf of Mexico." *SPE Drilling Engineering* (September 1988): 241–247.

M. Coudeyre, B. Le Goc, and J. P. Lombez. "Guiding a Horizontal Well Using Geological Monitoring." *World Oil* (July 1991): 59.

R. Dawson and P. R. Paslay. "Drillpipe Buckling in Inclined Holes" *Journal of Petroleum Technology* (October 1984): 1734–38.

J. A. Dech, et al. "New Tools Allow Medium-Radius Horizontal Drilling." *Oil & Gas Journal* (July 14, 1986): 95–99.

J. Dobson and T. C. Mondshine. "Unique Completion Fluid Suits Horizontal Wells" *Petroleum Engineer International* (September 1990): 42–48.

D. Eberts and R. D. Barnett. "Ultra High-Angle Wells Are Technical and Economic Success." *Oil & Gas Journal* (July 19, 1976): 115–120.

P. A. Edlund. "Application of Recently Developed Medium-Curvature Horizontal-Drilling Technology in the Sprayberry Trend Area." *Journal of Petroleum Technology* (September 1988): 1178–82.

R. W. Fincher. "Short-Radius Lateral Drilling: A Completion Alternative." *Petroleum Engineer International* (February 1987): 29–34.

L. H. Flack and W. C. Goins, Jr. "New Relief Well Technology Is Improving Blowout Control." *World Oil* (December 1983): 57–61.

J. D. Fultz and F. J. Pittard, *Openhole Drilling Using Coiled Tubing and a Positive Displacement Mud Motor.* SPE 20459. Society of Petroleum Engineers. New Orleans, LA, September 23–26, 1990.

R. L. Gates and G. Schwab. "Specialized Drilling Systems Set New World Records in High-Angle Holes." *Journal of Petroleum Technology* (February 1984): 241–248.

F. Giger. *Horizontal Well Production Techniques in Heterogeneous Reservoirs.* SPE 13710. Presented at the Society of Petroleum Engineers 1985 Middle East Oil Technical Conference and Exhibition. Bahrain, March 11–14, 1985.

F. Giger, L. Reiss, and A. Jourdan. *The Reservoir Engineering Aspects of Horizontal Drilling.* SPE 13024. Society of Petroleum Engineers. Houston, TX, September 16–19, 1984.

Giin-Fa Fuh, et al. *Borehole Stability Analysis for the Design of First Horizontal Well Drilled in the U.K.'s Southern "V" Field.* SPE 20408. Society of Petroleum Engineers. New Orleans, LA, September 23–26, 1990.

R. D. Grace, A. F. Kuckes, and J. Branton. "Operations at a Deep Relief Well." *World Oil* (May 1990): 44–54.

R. D. Grace and B. Cudd. "Fluid Dynamics Used to Kill South Louisiana Blowout." *World Oil* (April 1989): 47–50.

R. H. Holifield and B. Rehm, "Recompletion by Horizontal Drilling Pays Off." *World Oil* (March 1989): 42–43.

S. D Joshi. "Methods Calculate Area Drained by Horizontal Wells." *Oil & Gas Journal* (September 17, 1990): 77–80.

A. P. Jourdan, P. Armessen, and C. Mariotti. "Horizontal Well Operations—Part 4: Horizontal Drilling Has Negative and Positive Factors." *Oil & Gas Journal* (May 23, 1988): 37–40.

A. P. Jourdan and G. Baron. "Horizontal Well Proves Productivity Advantages." *Petroleum Engineer International* (October 1984): 23–25.

R. Jurgens, R. Bitto, and B. Henderson. "Horizontal Drilling and Completions: A Review of Available Technology—Part 1: Short- and Medium-Radius Horizontal Drilling." *Petroleum Engineer International* (February 1991): 14–21; and Part 2: "Medium- and Long-Radius Horizontal Drilling." *Petroleum Engineer International* (March 1991): 32–37.

H. Karlsson and R. Bitto. "Worldwide Experience Shows Horizontal Well Success." *World Oil* (March 1989): 51–56.

D. Kerr. "Designing Tangent Sections for Medium-Radius Horizontal Wells." *World Oil* (March 1991): 45–47.

D. Kerr. "How to Drill a Smooth Medium-Radius Well." *World Oil* (March 1990): 46–47.

W. J. Land and M. B. Jett. "High Expectations for Horizontal Drilling Becoming a Reality." *Oil & Gas Journal* (September 24, 1990): 70–79.

F. Leraand, et al. *Relief Well Planning and Drilling for a North Sea Underground Blowout.* SPE 20420. Society of Petroleum Engineers. New Orleans, LA, September 23–26, 1990.

B. M. Lowen and G. D. Gradeen. "Canadian Operator Succeeds in Slant-Hole Drilling Project." *Petroleum Engineer International* (August 1982): 40–52.

D. C. Loxam. "Texaco Canada Completes Unique Horizontal Drilling Program." *Petroleum Engineer International* (September 1982): 40–52.

A. Lubinski. "Maximum Permissible Dog-Legs in Rotary Boreholes." *Transactions of the American Institute of Mining and Metallurgical Engineers* 222 (1961): I–175.

B. J. Mahony. "Horizontal Drilling Use on the Rise: Why and How." *World Oil* (October 1988): 45–57.

T. Mall and R. Fincher. "Michigan Operator Salvages Well Using Lateral Drilling." *Oil & Gas Journal* (June 9, 1986): 33–38.

C. Mariotti and E. Kou. "Elf Improves Horizontal Drilling at Rospo Mare." *Petroleum Engineer International* (August 1988): 30–35.

W. H. McMillian. "Planning the Directional Well—A Calculation Method." *Journal of Petroleum Technology* (June 1981): 952–962.

R. McNally. "Horizontal Drilling Finding a Niche." *Petroleum Engineer International* (September 1990): 38–41.

K. K. Millheim. "Operators Have Much to Learn about Directional Drilling." Part One. *Oil & Gas Journal* (November 6, 1978): 63.

S. D. Moore. "High-Angle Drilling Comes of Age." *Petroleum Engineer International* (February 1987): 18–22.

S. D. Moore. "Meridian Oil Finds Success with Horizontal Wells." *Petroleum Engineer International* (November 1989): 17–22.

G. Moritis. "Horizontal Drilling Scores More Successes." *Oil & Gas Journal* (February 26, 1990): 53–64.

G. Moritis. "Worldwide Horizontal Drilling Surges." *Oil & Gas Journal* (February 27, 1989): 53–63.

T. M. Muhleman, Jr. "What's Happening in Drilling." *World Oil* (March 1986): 19.

F. G. D. Mullier. "Much Trouble Caused by Crooked Holes." *Oil Weekly* (April 19, 1924).

Oil & Gas Journal. "Horizontal Chalk Well Blowout Killed." *Oil & Gas Journal* (May 21, 1990): 22–23.

Oil & Gas Journal. "Pearsall Oil Well Completed with Dual Drainholes." *Oil & Gas Journal* (October 20, 1990): 37.

Oil & Gas Journal. "Petroleum 2000." *Oil & Gas Journal* (August 1977): 169.

G. A. Petzet. "Horizontal Drilling Fanning Out as Technology Advances and Flow Rates Jump." *Oil & Gas Journal* (April 23, 1990): 21–24.

G. A. Petzet. "Slant Holes Tap Shallow Gas under Lake." *Oil & Gas Journal* (May 14, 1984): 72–74.

M. M. Power, R. Chapman, and R. O'Neal. "Horizontal Well Set Depth Record." *Petroleum Engineer International* (November 1990): 36–38.

L. Ranney. "Drilling Wells Horizontally." *The Oil Weekly* (January 20, 1941).

B. Rehm. "Horizontal Drilling Applied in Slim Holes." *Petroleum Engineer International* (Feb 1987): 24–28.

L. H. Reiss and A. P. I. Jourdan. *Offshore and Onshore European Horizontal Wells.* OTC 4791. Presented at the 16th Annual Offshore Technology Conference. Houston, TX, May 7–9, 1964.

G. Renard and J. M. Dupuay. "Formation Damage Effects on Horizontal-Well Flow Efficiency." *Journal of Petroleum Technology* (July 1991): 786–869.

B. Sayers. "Capping Blowouts from Iran's 8-year War." Part 2. *World Oil* (July 1991): 81–82.

B. A. Sheikholeslami, et al. "Drilling and Production Aspects of Horizontal Wells in the Austin Chalk." Journal of Petroleum Technology (July 1991): 773–779.

M. C. Sheppard, C. Wick, and T. B. Burgess. "Designing Well Paths to Reduce Drag and Torque." *SPE Drilling Engineering* (December 1987): 344–350.

R. D. Sidman, J. Le Blanc, and B. Youngblood. "Quadratic Calculation Improves Interpretation of Directional Surveys." *Oil & Gas Journal* (January 23, 1978): 69–72.

M. Soliman, et al. "Planning Hydraulically Fractured Horizontal Completions." *World Oil* (September 1989): 54–58.

G. P Starley, et al. "Full-Field Stimulation for Planning and Reservoir Management at Kuparuk River Field." *Journal of Petroleum Technology* (August 1991): 974–982.

R. L. Stramp. *The Use of Horizontal Drain Holes in the Empire Abo Unit.* SPE 9221. Society of Petroleum Engineers. Dallas, TX, September 21–24, 1980.

J. Striegler. "Arco Finishes Fourth Horizontal Drainhole." *Oil & Gas Journal* (May 24, 1982): 55–61.

J. L. Thorogood and S. J. Sawaryn. "The Traveling-Cylinder Diagram: A Practical Tool for Collision Avoidance." *SPE Drilling Engineering* (March 1991): 31–36.

H. Uzcategui, D. Hewitt, and R. Golindano. "Precise Guidance Puts Record-Depth Relief Well on Target." *World Oil* (June 1991): 39ff.

H. J. Vrielink and A. M. Hippman. "The Optimization of Slant-Well Drilling in the Lindbergh Field." *SPE Drilling Engineering* (December 1989): 307–314.

T. M. Warren. "Directional Survey and Proximity Log Analysis of a Downhole Well Intersection." *Journal of Petroleum Technology* (December 1981): 2351–62.

R. C. Wilson and D. N. Willis. "Successful High-Angle Drilling in the Statfjord Field." SPE 15465. Society of Petroleum Engineers. New Orleans, LA, October 5–8, 1986.

J. Wu and H. C. Juvkam-Wold. "Drag and Torque Calculations for Horizontal Wells Simplified for Field Use." *Oil & Gas Journal* (April 29, 1991): 49–56.

CHAPTER 2

DRILLING TOOLS

SUMMARY

Drilling tools are equipment used downhole for drilling vertically, directionally, or horizontally, including special tools for measuring and guiding the direction and angle of the hole. Drillstring designs must have sufficient strength to support the drillstring weight, including an overpull factor for increased forces due to drilling stresses, excess drag and torque, and for sticking. Bottomhole assemblies provide bit weight, stabilization, and control the direction of the hole. Two basic classifications are rotary assemblies (rotated by the rotary or top drive) and motor assemblies, in which a positive displacement motor or turbine in the bottom part of the assembly rotates the bit. All assemblies use drill collars and bits and other tools depending upon the reason for using the assembly. Limber, reaming, hole-opening, coring, testing, and fishing rotary assemblies are used in vertical holes and for some directional applications. Stiff rotary or motor assemblies are used to maintain hole direction. Rotary assemblies change the hole direction up or down using stabilizers as fulcrums. Motor assemblies change the direction of the hole in any reasonable direction with bent subs and turbines or motors, sometimes, with bent housings. Measuring instruments record wellbore drift and direction and bit direction. Other data may be recorded to aid drilling of directional and horizontal holes.

DOWNHOLE EQUIPMENT

Downhole equipment includes all tools and instruments for drilling and related activities in vertical, directional, and horizontal holes. Much of the vertical drilling equipment is also used in both directional and horizontal drilling. Basic groups of downhole equipment include the drillpipe string, the bottomhole assembly, and deviating tools.

Tubulars and other similar equipment have various screw-type connections and different diameters for different hole sizes. Most tubulars are identified by the outside and sometimes inside diameters (ID), the grade of steel, and by their weight (lbs/ft). If only one diameter (size) is given, then it is the outside diameter (OD). This applies to tubes and connectors.

Tool joint connectors frequently have different outside and inside diameters compared to the body of the tube. Vendor catalogs, service company brochures, and other publications contain equipment data and specifications. Directional and horizontal holes are drilled with standard drilling rigs and equipment, although some rig modifications are required for special situations such as slant-hole drilling.

It is important to use the correct equipment and only that equipment necessary to do the job. It is generally better not to run any equipment in the hole that cannot be recovered by fishing procedures. While preparations must be made to fish out anything run in the hole, this should not deter running the required equipment (see Fig. 2–1).

DRILLPIPE STRING

The drillpipe string is the combined length of joints of drillpipe connected together and suspended in the hole. It transmits power from the surface to the bottomhole assembly for lifting, lowering, rotation, and other work. It serves as a conduit so drilling fluid (mud) can circulate down the hole inside the drillstring. The drillpipe string may be exposed to extremely harsh operating conditions. The severity depends upon the directional pattern and depth. The drillpipe string should be designed and operated within design specifications.

It is most common to have only one size of drillpipe in the drillstring. However, combination or tapered strings may contain two sizes of drillpipe with the larger, stronger pipe on top. They increase the maximum overpull. Tapered strings are seldom used except under conditions of high-loading in deeper holes. They require extra tools and increase trip time because of the need to change slips and elevators. An extra ram-type preventer provides for sealing the extra pipestring.

Figure 2–1
Drill tools

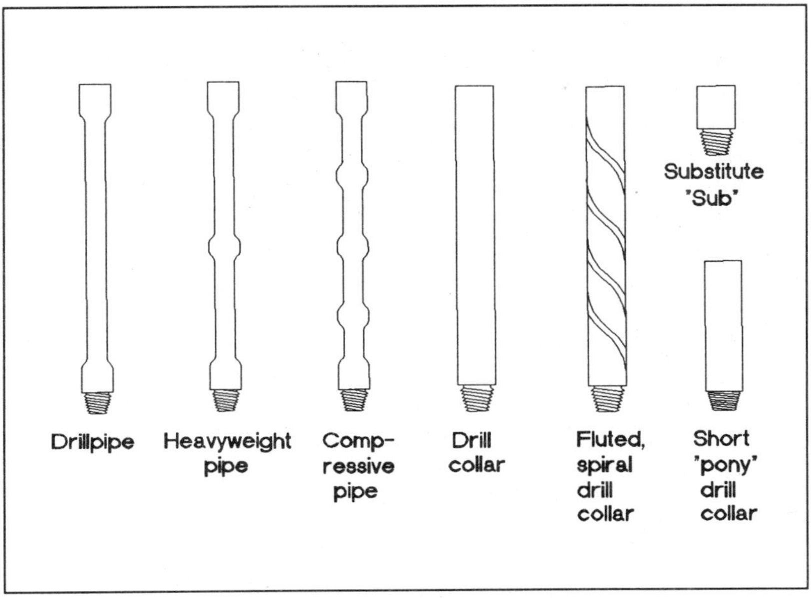

The screw-type tool joint connections on the end of the drillpipe are stronger than the pipe body and seldom fail due to tensional pull. They must be tightened to the correct torque. A loose joint will wobble off, and a too-tight one may crack the box or pin. Either causes a fishing situation.

All used drillpipe is not necessarily the same, even with a history of the same number of drilled feet. Some is in poor condition due to normal wear, slip cutting, and improper tool joint makeup. Severe service causes fatigue that is cumulative and difficult to detect. Service in H_2S environments causes embrittlement. It is important to know the history of the drillpipe and have good tubular testing practices and procedures.

DRILLPIPE TYPES

Drillpipe is heavyweight, high-grade steel pipe with butt shoulder, screw-type connections on each end. It supports its weight, the weight of the bottomhole assembly, and any additional torque, drag, or sticking forces encountered. The drillpipe transmits rotational force or torque for drilling and is a conduit for the mud. Sizes range from 2 7/8 in. to 6 5/8 in.; the most common are 4 1/2 in. and 5 in. Steel grades include D, E, G, and S-135 with minimum yield strengths of 59 Mlb/in^2 (thousands of lbs per square inch) to 135 Mlb/in^2, and special grades.

ALUMINUM DRILLPIPE

Aluminum drillpipe has a favorable combination of low weight and high strength. It can reduce assembly weight appreciably, sometimes approaching 50%. Excluding other factors, this is an appreciable increased depth rating. Aluminum drillpipe has been tested somewhat extensively because of these favorable characteristics. However, operational problems with aluminum drillpipe generally outweigh the benefits. Tool joint and body wear are excessive. Aluminum pipe is more flexible than steel pipe, so it is subject to more buckling and impact damage. It is more difficult to fish for or with aluminum pipe, and it corrodes under many normal drilling conditions. Aluminum drillpipe is seldom used for these and similar reasons. It still is considered because of its favorable weight and strength properties.

COILED TUBING

Coiled tubing is a small diameter, long, high-strength, ductile steel tube available in various diameters and strength ratings. Lengths of 10,000 – 20,000+ ft are coiled on a large trailer-mounted reel. The tubing unwinds and passes through an injection head into the wellbore. The head lifts and lowers the tubing while providing a pressure seal. Coiled tubing carries logging and perforating tools in directional and horizontal holes. It serves for low-volume acid cleanout, pressure control, squeezing, and other work inside the pipe. Working with coiled tubing often is a high-risk operation, and the risk increases with depth and increasing hole angles.

PRODUCTION TUBING

Production tubing is small diameter pipe used primarily in completions. It contains the oil and gas as they flow upward through the wellbore to the surface. Sizes range from 1 1/16 in. to about 5 in.; the most common sizes are 2 3/8 in. and 2 7/8 in. Drill tubing has stronger connections and replaces regular tubing for conducting operations in smaller diameter holes and in completions and workovers.

BOTTOMHOLE ASSEMBLY

The bottomhole assembly (BHA) includes all regular drilling tools connected to the bottom of the drillpipe string. The purpose and design of this equipment is described in detail later in this chapter.

DRILL COLLARS

Drill collars are similar to drillpipe but heavier, with a thicker wall and stronger connections. Heavier construction maintains integrity under the high and fluctuating stresses that occur near the bottom of the hole. Drill collars provide bit weight and rigidity or flexibility as needed, so that different assembly types operate as designed. They also maintain the "free point" within the stronger drill collar assembly in most cases.

Fluted or spiral drill collars are similar and serve as regular collars. A recessed, spiral-shaped section reduces wall contact area to prevent wall sticking and serves as a channel for mud. Nonmagnetic drill collars serve the same purpose as regular drill collars. They are constructed from nonmagnetic material so tools inside the drill collar can record compass measurements. Pony drill collars are shorter than standard drill collars and constructed of regular or nonmagnetic steel. They permit spacing other tools correctly on the bottomhole assembly. Substitutes or subs are short pony drill collars (about 4 ft or shorter).

Table 2–1
Drill Collar and Hole Sizes.

Drill Collar Diameter (OD), in.	Hole (ID), in.
4 3/4 to 5	6 1/8 to 6 3/4
5 1/2 to 6	7 to 7 3/4
6 1/2 to 7	8 1/2 to 8 3/4
7 1/2 to 8	9 7/8 to 10 3/4
8 to 10	12 1/4 and larger

HEAVY DRILLPIPE

Heavy drillpipe is similar to regular drillpipe except that it is heavier and has an enlarged section shaped like a tooljoint near the center for added lateral support. It replaces drill collars, reducing drillstring weight.

COMPRESSION PIPE

Compression pipe is similar to heavy drillpipe except that it may be made of higher grade steel and have two or more tooljoint-shaped center sections. These provide additional lateral support, distribute bending stresses, and protect the pipe body from wear against the wall of the hole.

STABILIZERS

Stabilizers are similar in appearance to a sub but have short, heavy blades on the circumference. They provide stabilization on the bottomhole assembly for directional and horizontal assemblies. Various types and blade configurations include fixed (spiral or straight-blade), replaceable straight-blade, short-blade, near-bit, and swiveling body types. Some are available in both regular and nonmagnetic steel. Adjustable blade stabilizers have adjustable blades to increase or decrease diameter. One type has button inserts that expand hydraulically by the drilling fluid. The blades expand for stabilization and retract when not needed (see Fig. 2–2).

Field practice has either not determined which is the best tool for a specific application, or it has not been reported. It probably is site-specific. There are certain theoretical, practical reasons for using certain types. Larger contact areas reduce embedment in softer formations. Smaller contact areas may reduce drag in harder formations.

Replaceable roller cutter types are efficient, especially in medium to hard formations; however, they may tend to ball up or become coated with formation cuttings in softer formations. They have a large body that is very difficult to mill over if a failure occurs.

Fixed, straight-blade stabilizers commonly have wide, "softer" steel blades dressed with a hard facing material, such as tungsten carbide, on the wear surface. If required by a fishing job, the blades can be mill cut through the softer metal between the hard facing material and the tool body with minimum difficulty.

Theoretically, a curved-blade stabilizer minimizes impact forces on the drill collar assembly and, more importantly, on the adjacent connectors as compared to fixed, straight-blade stabilizers. This reduces the risk of fatigue failures in the connections. Curved-blade stabilizers have a larger wall contact surface area compared to straight-blade tools with the same length and blade width. Swiveling body stabilizers are not widely used.

REAMERS

Reamers are similar in appearance to stabilizers. They open or ream an undergauge hole to original size and otherwise smooth the wall of the hole. Regular reamers are placed in the bottomhole assembly. String reamers are placed in the drillpipe string for reaming upper hole sections while drilling. The terms reamer and stabilizer often are interchangeable, since frequently either tool may be used for the same purpose (see Fig. 2–3).

Figure 2–2
Stabilizers
(courtesy of Eastman Christensen, a Baker-Hughes company)

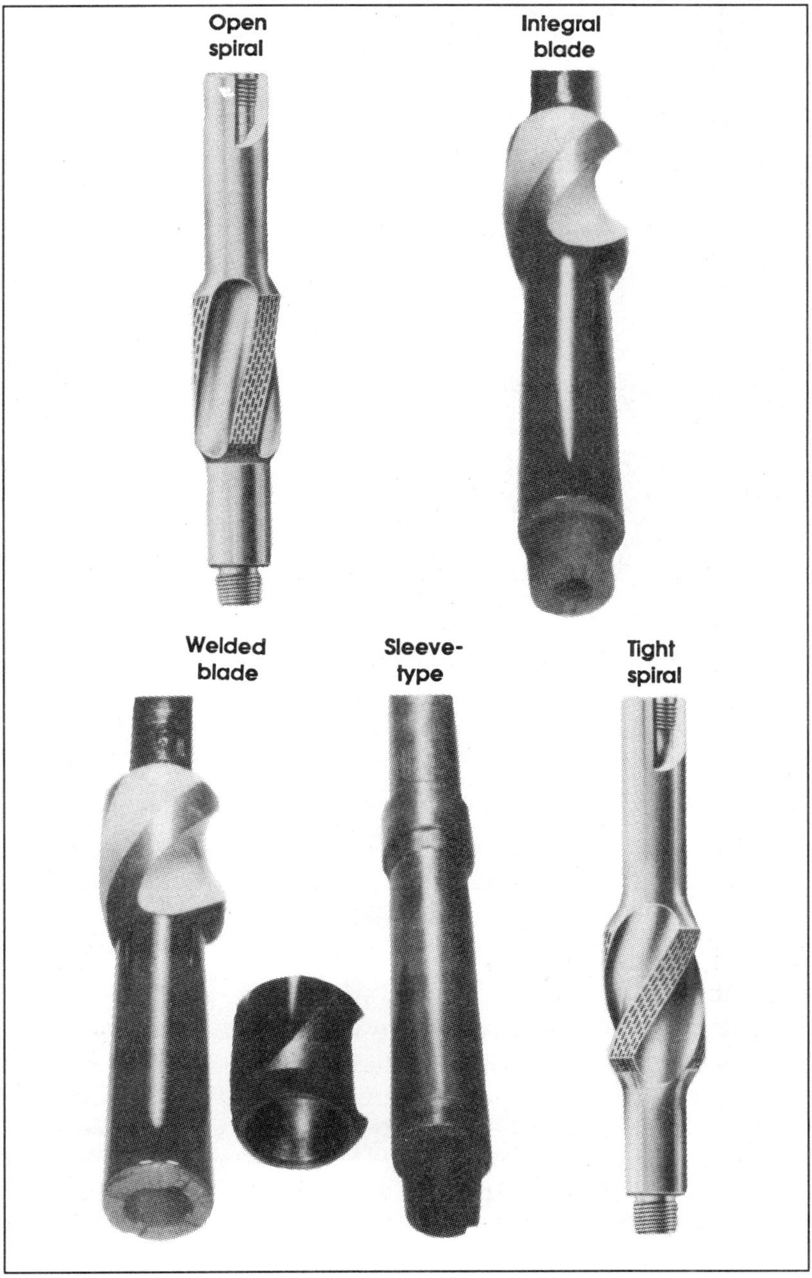

Figure 2–3
Reamers
(courtesy of Eastman Christensen, a Baker-Hughes company)

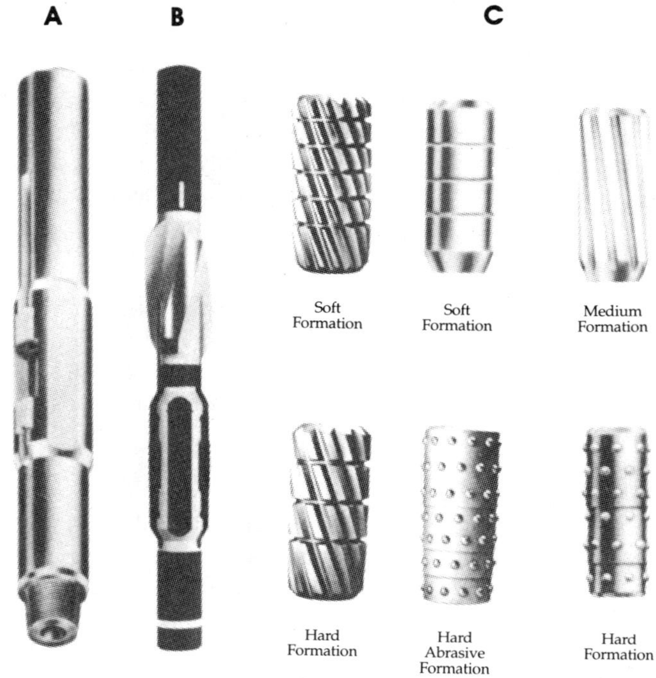

A. Roller Reamer; B. Combination Roller Reamer and Stabilizer; C. Roller Reamer Cutters

JAR-BUMPER SUBS

Jar-bumper subs help release stuck pipe by delivering a strong jarring blow in an upward or downward direction. The strength of the blow should be adjusted based on the type of tool and the manner of operation. Various tools are available, and they should be positioned in the bottomhole assembly below the top 3 to 5 drill collars. These collars supply weight for the jarring blow. The number of collars above the jar-bumper depends upon its specifications. Jar-bumper subs are very effective for releasing stuck tools and should be included in most drillstrings. There are both drilling and fishing jar-bumper subs. The drilling jar-bumper is used during drilling, and the somewhat stronger fishing jar-bumper is used for fishing operations (see Fig. 2–4).

KEYSEAT WIPER

Keyseat wipers connect on top of the bottomhole assembly and remove keyseats during drilling and tripping. A tapered-rib, sleeved, bottom-clutch type is best. The upper rib diameter should be the same as the drillpipe tool joint diameter, and the lower rib diameter should equal the diameter of the largest drill collar. If it becomes stuck, it can be jarred down and out of a keyseat. This preventive tool can provide a quick means of releasing stuck tools in many cases.

SHOCK SUBS

Shock subs reduce bit bounce, which helps the bit remain in contact with the formation face on the bottom of the hole. They also help reduce vibration in the bottomhole assembly. These should not be used on assemblies in which they cause decreased assembly stiffness and reduce efficiency.

DEVIATING TOOLS

Deviating tools fit on the bottomhole assembly to allow the assembly to serve a different purpose, such as changing or maintaining the direction or angle of the drill hole. Their design and purpose is described here and also in the section about bottomhole assemblies later in this chapter.

TURBINES AND MOTORS

Turbines and positive displacement motors (PDMs or mud motors) use the pressure and volume of the circulating mud to rotate the bit. This, in conjunction with other tools, provides an efficient method to change the hole direction. Positive displacement motors are used more commonly in directional and horizontal drilling. This is attributed in part to lower hydraulic horsepower requirements, a wider range of sizes, a better selection of bits, lower unit cost, and flexibility of rotational speeds and torque combinations. This text generally refers to drilling with motors as a convenience. Usually either a positive displacement motor or turbine can be used, depending upon the pumping equipment and specific well conditions.

Turbines are available in different sizes, but the minimum size is about 5 in. in diameter, so they cannot be used in small diameter

Figure 2–4
Drilling jar
(courtesy of Eastman Christensen, a Baker-Hughes company)

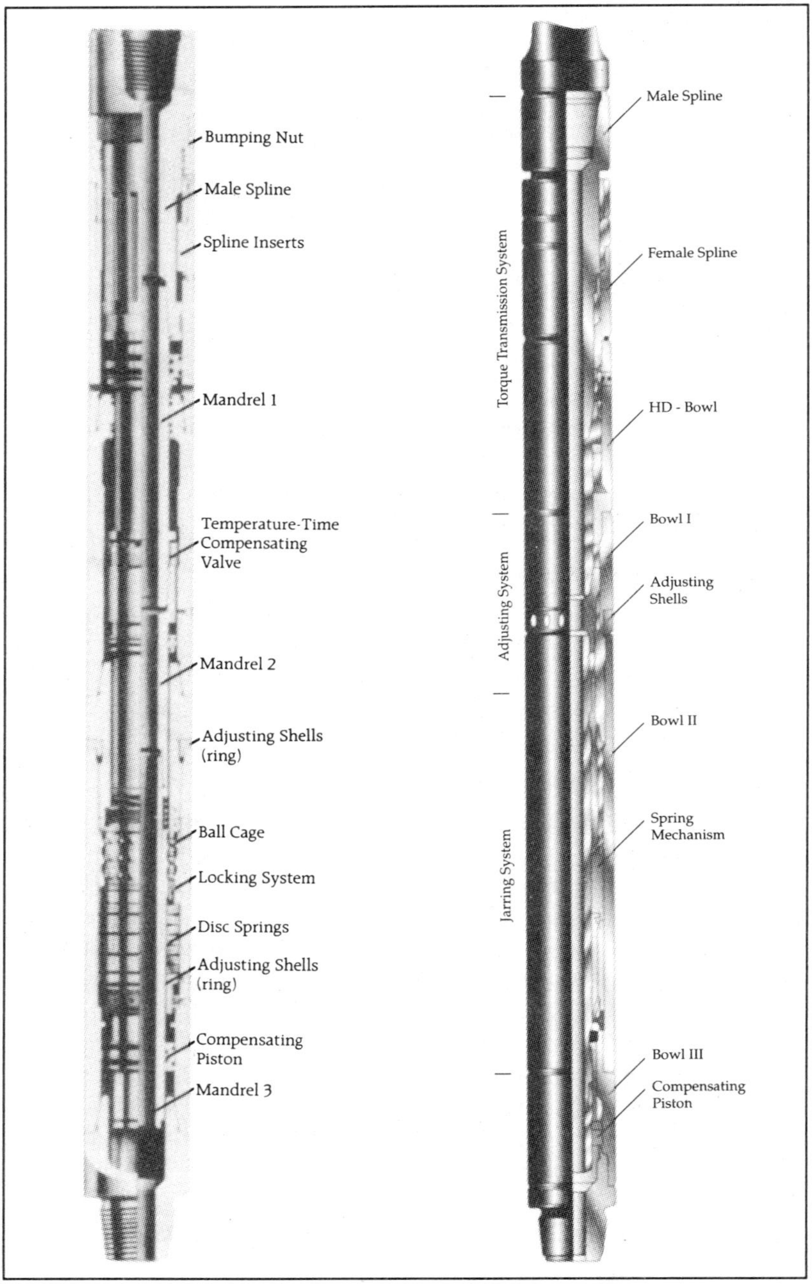

holes. They have a set of stationary stator vanes connected to the housing. These deflect mud against the vanes on the rotor, rotating the drive shaft and bit connected to its lower end. Each stator and rotor-vane combination is a turbine stage. Multiple stages increase turbine power. Turbines have from about 70 to 150 stages depending upon the size and use of the turbine (see Fig. 2–5).

Turbines usually operate at higher rotational speeds, than positive displacement motors in the range of 1,000 revolutions per minute (rpm). Bit selection is more restricted for turbine drilling. Solid-bodied bits are more common because of the high rotational speed. Turbines generally require higher hydraulic horsepower. This may account for their increased usage offshore, since marine rigs frequently have excess pump capacity. Turbine modifications for directional drilling include the offset turbine with twin stabilizing blades or similar offsetting devices for directional drilling and guided turbodrills.

Positive displacement motors are available in a wide range of sizes from slightly less than 2 in. to more than 9 in. in diameter. They have a sinusoidal-shaped rotor fitted inside the stator, an elongated, rubber-lined cavity. The rotor has one or more lobes and is located inside a stator that has one more lobe than the rotor. Common motors use one rotor and two lobes for high torque. Increasing the number of lobes increases speed and reduces torque for a given size. Mud passing through the cavity turns the rotor that connects to and rotates a drill bit (see Fig. 2–6). Liquid mud rotates most motors. A few have been modified for operation by air, although this is seldom used. Motors have a wide range of speeds from about 100 rpm to more than 800 rpm. The most common operational speeds vary from about 150–300 rpm. There are a wide range of drill bits (including roller bits) available for these operational speeds.

One popular motor variation is the bent-housing motor, which has a bend constructed near the lower end. A universal joint transmits power through the bent section. This serves as a primary deflection tool for deviating. A deflection pad on the base of the bend reduces wear on the housing. It also increases the lateral force on the bit to increase the rate of angle buildup. Some motors have adjustable pad thicknesses for changing the angle-build rate. Other versions use a pad on the lower end of the housing. Another variation has two stabilizer blades in a V shape on the lower housing. Bent-housing motors are an efficient, commonly used deviating tool.

Another PDM variation is the double-joint motor, which has two bends in opposite directions. This increases effective bend angle

Figure 2–5
Turbine
(courtesy of Drilex Systems, Inc.)

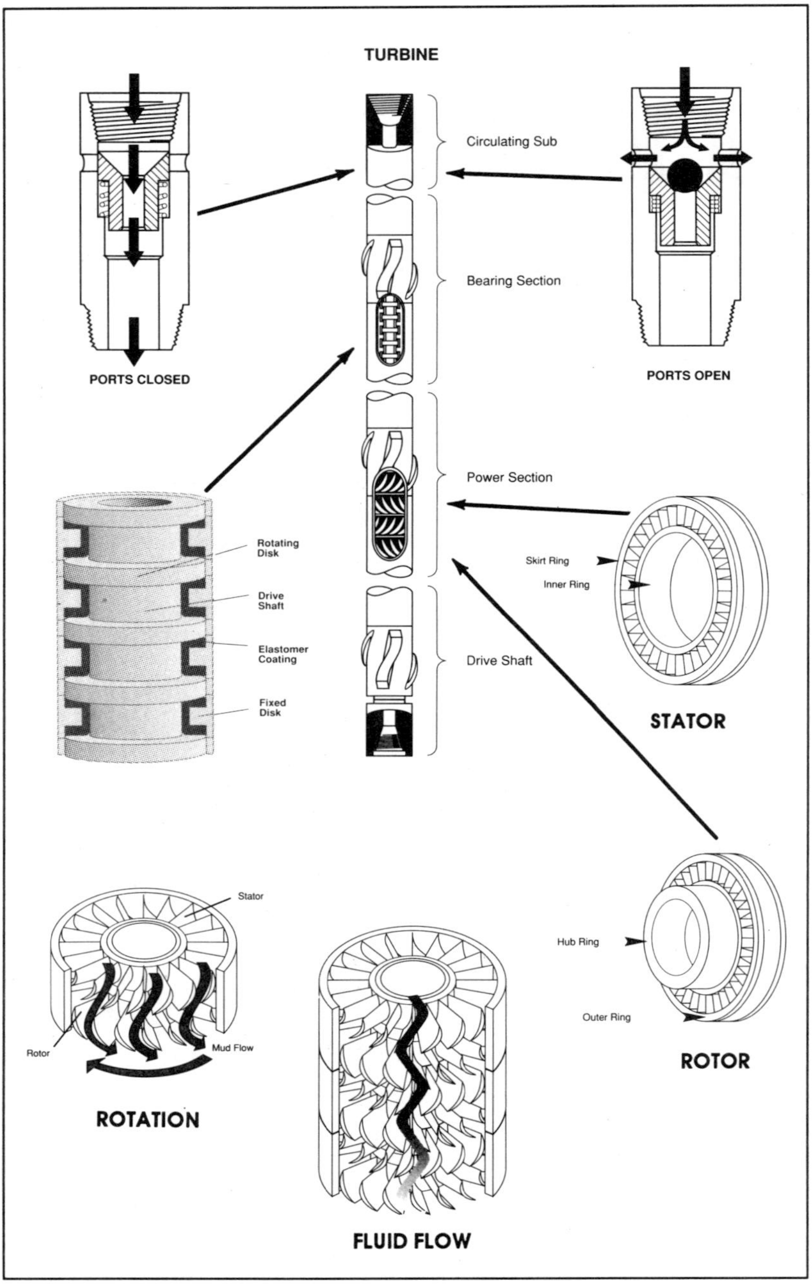

Figure 2–6
Positive displacement motors
(courtesy of Eastman Christensen, a Baker-Hughes company)

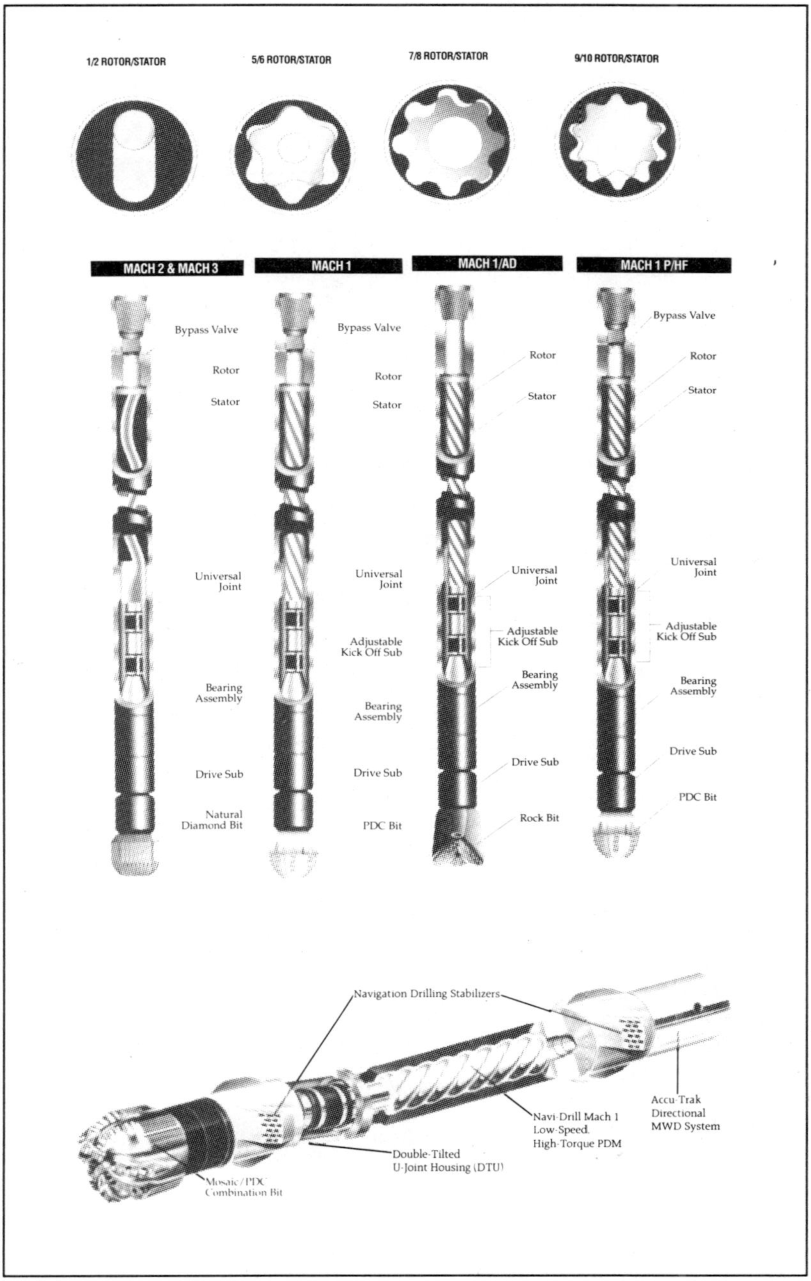

and reduces tool length building angle at a higher rate. The tool has a short bit offset the horizontal distance between the center line of the bit and the centerline of the upper hole (See the discussion of bent subs in the section Horizontal Applications later in this chapter). Sometimes the drillstring rotates for drilling straight ahead, or the motor rotates the bit for directional drilling (see Fig. 2–7).

BENT SUB

Bent subs are primary deviating tools positioned above motors and turbines in rotary assemblies. It is a standard sub, modified so the bottom connection points in an off-centered direction relative to the axis of the sub body and upper drilling assembly. This creates a side force and deflects the motor and drill tools (connected to the bottom of the sub) in the direction of the off-centered connection.

Figure 2–7
Positive displacement motor types

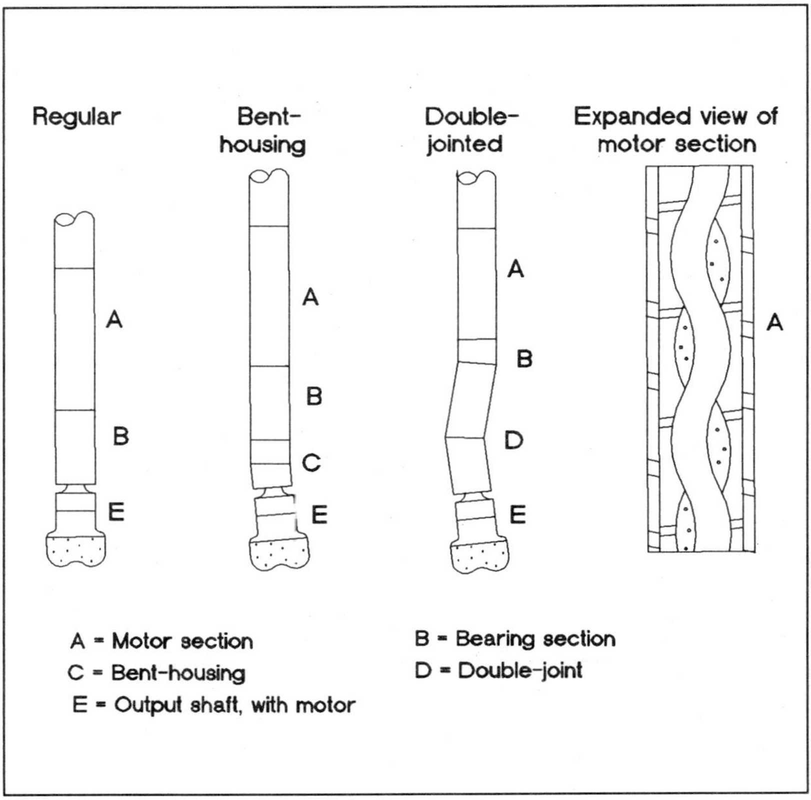

Bent subs are identified by the angle of deflection, which ranges from 1.5° to more than 3°. Adjustable bent subs are similar to fixed-angle bent subs, except the bend angle is adjustable while drilling. This saves tripping time to replace the fixed-angle bent sub when a different size of sub is needed.

FLEXIBLE JOINTS

Flexible joints or knuckle joints are similar to bent subs except that the tool has a "zero" bend angle for tripping. The hydraulic force of the mud causes the tool to shift to a fixed-bend angle, normally between 1.5°–3°. In this position, the tools act similarly to regular bent subs. A modified version of the tool provides for changing angle mechanically with wedges run on a wireline. This tool is seldom used.

DRAIN SUB

Drain or circulating subs allow drilling fluid to drain or flow out of the drillstring when pulling the drillstring out of the hole. This prevents the drilling fluid from overflowing and spilling onto the surface as each stand is disconnected. This could create a potentially hazardous working condition for operating personnel, as well as losing expensive drilling fluid. A common version has a bypass that opens by dropping a ball and circulating it to bottom. The tool may be incorporated in the construction of some turbines and positive displacement motors.

WHIPSTOCKS

Whipstocks were the first reliable deviating tool, but now they have been replaced by the more efficient bent sub and mud motor or turbine deviating systems. Modified whipstocks are used for deviating in cased holes and for short-radius horizontal drilling systems. It has a tapered body that guides directional tools away from the axis of the wellbore.

JET SUBS

Jet subs constantly bypass part of the drilling fluid so that it does not pass through the motor or turbine. Cleaning the hole adequately may require large volumes of mud that may overpower the turbine or motor. A jet sub positioned above the turbine or motor bypasses part of the mud directly into the annulus. This provides higher mud volumes for hole cleaning without damaging the turbine or motor. For example, a specific hole situation may need a 125 gallons per minute (GPM) flow rate for sufficient cleaning, whereas the motor only needs 100 GPM. The jet sub would be sized to bypass 25 GPM; the bit jet nozzles would also be sized for the 100 GPM flow rate.

MULESHOES

The muleshoe slot and lug allows the positioning of measuring instruments in the bottomhole assembly (BHA). A muleshoe orienting sub connected in the BHA contains an internal lug or key. The key is positioned so that it is in a fixed position relative to the bit face when the muleshoe orienting sub is connected to the deviating assembly. The measuring instrument carrier has a muleshoe slot on bottom. The carrier is lowered into the hole on a single-strand wireline. It turns automatically as it lands in the measurement sub below the nonmagnetic collar. The muleshoe slips over the key, positioning the carrier in a fixed position relative to the bit face.

DRILLSTRING

The drillstring includes all the equipment suspended in the hole, such as the drillpipe string, the bottomhole assembly, and the deviation equipment. Drillstrings may be exposed to extremely harsh operating conditions, especially in directional drilling and even more so in horizontal drilling. The severity depends upon the directional pattern and depth. Experience and information from similar wells in the area are very helpful in the selection of design criteria.

WEIGHT AND BUOYANCY

Drillstring weight is the weight of the drill tools suspended in the hole measured in thousands of pounds (Mlbs). Air weight is the weight of the drillstring in air, normally used as the basis for calculations. Buoyant weight is the drillstring weight suspended in drilling fluid, normally the weight shown on the rig weight indicator. This is less than the air weight and depends upon the density of the fluid. It is the total load supported by the mast.

The more significant weight is that on the top joint of drillpipe. This is usually the total load or buoyant weight less the weight of the traveling block and other equipment above the top joint of drill pipe. This averages about 12,000 lbs for smaller rigs and 15,000–20,000 lbs for larger rigs. The following example is based on the load on the top joint of the drillpipe.

EXAMPLE 2.1:

Ten thousand feet of 4 1/2 in., 16.60 lbs/ft drillpipe has an air weight calculated by:

Air weight of the drillpipe = (pipe length in ft/1000) x (pipe weight/ft)
= (10,000/1000) x (16.60)
= 160 Mlbs (160,000 lbs).

The weight of a common 7 in. bottomhole assembly with about 350 ft of drill collars weighing 100 lbs/ft would be calculated by:

Air weight of drill collars = (collar length (ft)/1000) x (collar weight/ ft
= (350/1000) x (100)
= 35 Mlbs (35,000 lbs).

Air weight of the total string = (air weight of the drillpipe) + (air weight of drill collars)
= 160 + 35 =195 Mlbs

This is the air weight or the total weight if the well used air for drilling fluid; buoyancy due to mud must be deducted from the air weight. A 12 lbs/gal drilling fluid has a buoyancy factor of 0.8166. Therefore, the buoyant weight or true weight on the top joint of drillpipe, rounded to the nearest thousand pounds, is calculated by:

Buoyant weight of the total string = (air weight of the total string) x (buoyancy factor)
= (195) x (0.8166)
=159 Mlbs (159,000 lbs).

This is the pipe weight or hook load registered on the rig weight indicator. To be exact, the weight indicator also will show the weight of the traveling block and tools as described above.

All references to weights of drill tools commonly refer to the buoyant weight unless otherwise specified. Pipe weight specifications such as 14.5 lbs/ft refer to air weight but carry the designation of lbs/ft, sometimes abbreviated to lbs.

OVERPULL

Overpull is a measure of the amount of pull or loading on the drillstring over normal buoyant weight. It is an important criterion in drillstring design since it is the limiting lifting force (pull) that can be applied to the drillstring, such as during drag and sticking. Higher pulls damage the drillstring and may cause a failure. Overpull is the difference between the maximum safe lifting force applied to the top joint of drillpipe and the buoyant weight of downhole tools at total depth. It is based on the maximum tensile strength of the drillpipe.

A drillpipe hanging freely in the hole stretches due to its weight and the weight of the bottomhole drilling assembly. This normally is between 0.5–1.5 ft per 1,000 ft of drillpipe with an average size of bottomhole assembly. Drillpipe that is not overstressed returns to its normal length when the load is removed. The maximum tensile strength is the maximum loading the drillpipe will sustain before it becomes permanently deformed ("stretched") and will not return to its original length after removing the load. Overloading leads to ultimate failure, causing fishing jobs or related problems. This condition is difficult to detect but very important to the integrity of the drillstring.

Stronger drillstring limits must be designed for drilling complex directional drilling designs, multiple bends and turns, long deviated hole sections, and areas with known formation problems.

EXAMPLE 2.2:

Continuing with the data from Example 2.1, a 4 1/2 in., 16.60 lbs/ft Grade E drillpipe has a maximum tensile strength of 331 Mlbs. Therefore the maximum overpull would be calculated by:

Maximum overpull = (maximum tensile strength) –
(buoyant weight)
= (331 Mlbs) – (159 Mlbs)
=172 Mlbs (172,000 lbs).

Total tensile force on the top joint of drillpipe is 331 Mlbs with an overpull of 172 Mlbs, the maximum overpull load before failure. A safety factor should be used with overpull load calculations since failure at this point is imminent. Common safety factors for used drillpipe are about 80% of the maximum tensile strength and strongly dependent upon the condition of the pipe. Therefore, the maximum safe overpull is calculated by:

Safe overpull = (maximum strength x safety factor) –
(buoyant weight)
= (331 MlbsX 0.80) – (159 Mlbs)
= 106 Mlbs (106,000 lbs).

This is the safe overpull, and the total tensile force on the top joint of drillpipe is 264 Mlbs. The maximum safe overpull should not be used without trying various fishing techniques and other actions. Each company has (or should have) standards for recommended overpull.

Table 2–2
Recommended Overpull Values.

Well Depth Range, ft	Overpull, Mlbs
Less than 8,000	100
8,000 to 12,000	125
12,000 to 15,000	150
15,000 to 18,000	175
More than 18,000	200

Recommended overpull values are given in Table 2–2. They are slightly higher than the industry standards and are intended as guidelines, subject to well pattern complexity and design requirements. Pattern complexity and well conditions affect selection of overpull values. For example, a straight, vertical hole is designed for less overpull than a deep, extended-reach well with higher drag and torque.

FREE POINT

The free point is a neutral point, usually in the bottomhole assembly (BHA), that is neither in tension or compression. The free point concept is important in assembly design and operation. The BHA is subject to high torque and tensile stresses during drilling, especially in sections under compression or below the free point.

The free point should be maintained in the stronger drill collar assembly in regular vertical and directional drilling and in horizontal drilling when possible. There may be a problem in high-angle and horizontal drilling in this respect because of the difficulty of maintaining bit weight. Damage at the free point may be strongly dependent upon drillstring rotation. Apparently, fewer problems occur in high-angle drilling with a stationary drillstring with the bit rotated by a motor.

For example, if the free point is at the very bottom of an assembly suspended off bottom, and the entire drillstring is in tension. However, if all the drillstring weight is set on bottom (this is not normally done), the freepoint is at the surface and the entire drillstring is in compression.

EXAMPLE 2.3:

Assume that the assembly is lowered so that the bit exerts 20,000 lbs of weight on the bottom (a normal drilling situation). The free point is located in the BHA at a distance above the bottom of the assembly equivalent to 20 Mlbs of BHA weight. The drill collars (from Example 2.1) weighed 100 lbs/ft in air. The buoyant weight in 12 PPG mud is:

(100 lbs/ft) x (0.8166) = 81.66 lbs/ft.

This represents a drill collar length of:

(20,000 lbs of bit weight)/(81.66 lbs/ft) = 245 ft.

The free point is 245 ft above the bit. The distance from the top of the bottomhole assembly is:

(350 ft assembly length) – (245 ft in compression) =105ft.

Therefore, the bottom 245 ft of the BHA are in compression and the top 105 ft in tension. Assemblies should be designed so that the free point is in the top 20% of the assembly during normal vertical and directional drilling. In this case it results in a bit weight of 23 Mlbs.

DIRECTIONAL CONTROL

Overcoming the force of gravity is a fundamental problem in directional and horizontal drilling. Drillstrings have a very small diameter compared to their length. They are very limber considering their diameter, length, and weight. The bottomhole assembly (BHA) is a heavy weight hanging on the bottom of the drillstring, all suspended from the surface. This hangs vertically downward due to gravity and drills the hole in the same vertical direction. The BHA must overcome the force of gravity with a strong side force for directional drilling. The force is applied with stabilization, fulcrums, and operating techniques.

The limber rotary assembly, consisting of drill collars and a bit, drills vertically downward. Its performance is strongly affected by formation and operating conditions. A stiff, rigid rotary or motor assembly, sometimes called a hold assembly, is a common directional assembly. It drills a straight hole, vertically or at an angle, subject to the tendency of some formations to cause the hole to deviate. The stiff, rigid assembly fits closely in the hole, held in place by multiple stabilizers. The hole behind the bit confines the assembly. Rigidity and stiffness force the BHA to remain in the same relative position and conform to the direction of the centerline of the hole immediately above bottom. This points the bit so that it continues drilling in the same direction as the hole behind it. Efficiency increases with increasing stiffness and rigidity. Therefore, a stiff, rigid assembly follows the direction of the hole behind the bit closely. A less rigid assembly allows natural forces to exert more influence on the direction of the hole.

Drill collar stiffness increases with increasing collar diameter, so large diameter collars are more rigid than those with smaller diameters. However, there is a practical limit to the collar size that can pass freely in a given hole diameter after allowing for unrestricted movement of tools and drilling fluid. Stabilizers can increase the apparent rigidity of smaller collars, giving the BHA an effective stiffness approaching that of larger diameter collars. Two or more stabilizers positioned in the lower section of the BHA support it laterally with multiple contact points against the wall of the hole. Apparent rigidity also increases with increasing rotational speed, so operating a stiff BHA at higher rotational speeds increases efficiency (see Fig. 2–8).

Drill collars suspended in a position other than vertical bend and sag downward at a point above the bit due to their weight and the force of gravity. Bit weight applied by the drill collars located higher in the BHA causes additional bending due to column loading. The combined actions cause the collars to touch the side of the hole at the point of tangency, some distance above the bit. The distance between the bit and tangent point depends upon collar and hole sizes, inclination, and bit weight (see Fig. 2–8).

An angle-building rotary assembly can be constructed by placing a stabilizer between the bit and point of tangency (see Fig. 2–8). The stabilizer normally is positioned near the bit as a near-bit stabilizer. The stabilizer acts as a fulcrum. The weight of the bending drill collars above the stabilizer causes the lower end of the collars to pivot at the fulcrum stabilizer. This points the bit so that it drills in the upward direction. The stabilizer slides on the lower side of the hole with very little cutting action so the angle buildup rotary assembly drills the hole in a smoothly curved upward direction.

The rate at which the assembly builds angle depends on the size of the drill collars, bit weight, and rotary speed. A smaller collar on bottom increases the build rate. Higher bit weight increases angle-building action by columnar loading. Higher rotary speed reduces the angle-build rate because it increases the apparent stiffness of the assembly. The build rate can be adjusted by changing the distance between the stabilizer and bit a small amount with a short sub. This increases the leverage, so the BHA builds angle faster. The bit alone can act as an angle-building assembly, but the BHA is more effective with the near-bit stabilizer. It is possible to change the BHA to an angle-building motor assembly by placing a turbine or motor above the bit.

Placing a stabilizer above the point of tangency causes a reverse action, and the BHA becomes an angle-dropping rotary assembly. The stabilizer again acts as a fulcrum so the drill collars pivot at this point. The section of drill collars below the stabilizer bends and sags downward due to the pull of gravity, somewhat like the action of a pendulum (see Fig. 2–8). This angle-dropping rotary assembly drops angle by drilling the hole in a smoothly curved downward direction. Normally the stabilizer is placed about 60 ft above the bit. The exact position depends upon drill collar size and weight, hole diameter, inclination, and bit weight. The distance is adjusted with subs and pony drill collars to increase or decrease the rate of angle drop. The rate of drop can be increased by reducing bit weight and rotational speed. Angle-dropping assemblies are efficient, especially in vertical drilling where formations cause crooked or deviated holes. A turbine or motor is placed above the bit to make an angle-dropping motor assembly. These are seldom used except under conditions requiring horizontal control, because the regular angle-dropping assembly is highly efficient and is a safe tool to run. Rotary assemblies control the vertical and not the horizontal direction of the deviated hole.

Deviation motor assemblies have a bent sub positioned above a motor and bit. The bent sub serves as a fulcrum similarly to a stabilizer but with several significant differences. The bend of the bent sub is fixed in one position on the assembly. It forces the bit away from the centerline of the original hole in a direction opposite the apex of the bend angle. The drillstring does not rotate, only the bit. The bit, turned by the motor, drills a curving hole in the direction of the bit face or opposite the apex of the bend angle. The degrees of bend in the bent sub or bent housing control the rate of change of angle. The assembly can be turned to point the bit face in a different direction, and the bit drills in the new direction. The assembly can be turned for building or dropping angle or to change the hole direction to the right or left, or a combination of these. Assembly function does not depend upon gravity action.

There are various modifications of the deviation motor assembly. The bent sub may be replaced with an adjustable bent sub to change the angle during drilling. The bent sub and motor may be

Figure 2–8
Change vertical angle with rotary assemblies

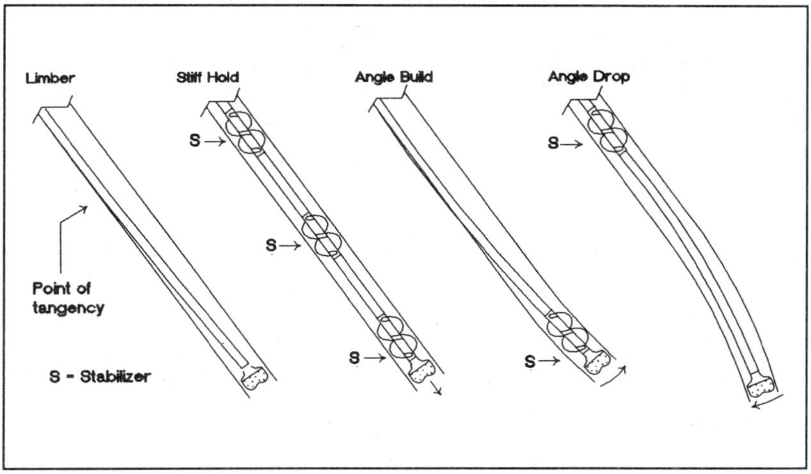

replaced with a bent-housing motor for changing hole direction similarly. A combination of turning tools such as a bent sub above a motor with a bent housing is more aggressive and changes angle faster (see Fig. 2–9).

A common method of describing stabilizer placement on bottomhole assemblies is by "position." Each position represents an assumed drill collar length of 30 ft, measured upward from the bit. A stabilizer immediately above the bit is identified by either "position 0" or "near-bit." Stabilizers at "positions 0 and 90" would be on top of the bit (or near-bit), and on top of the third drill collar counting upward from the bit. Stabilizers at "positions 60 and 90" would be on top of the second and third drill collars. Stabilizers at "positions 0, 2 and 60" would be on top of the bit (near-bit), on top of a 2-ft sub above the bit, and on top of the second drill collar. (Note that the stabilizer listed at "position 60" is NOT listed at "position 62.")

The stiff or hold assembly described above has stabilizer placements, listed in the order of increasing efficiency, as follows:

Position 0 and 30	Very inefficient
Position 0, 30, and 60	
Position 0, 30, 60, and 90	Most common
Position 0, 2, 30, 60, and 90	
Position 0, 2, 10, 30, 60, and 90	Very aggressive

Figure 2–9
Motor deviation sections

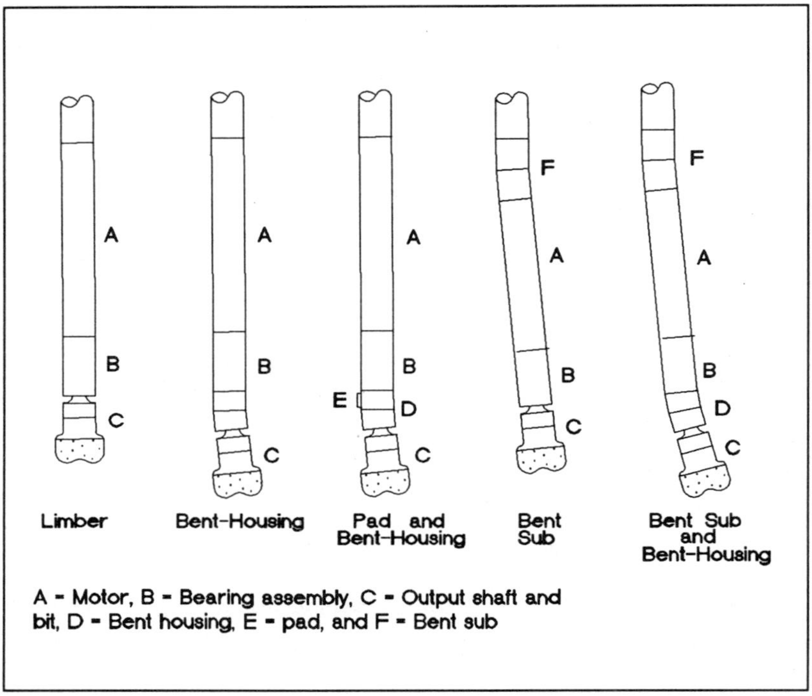

There is a risk of sticking when running an aggressive assembly in a deviated hole as described in Chapter 4.

BOTTOMHOLE ASSEMBLY

Bottomhole assemblies (BHA) include all drilling equipment connected to the bottom of the drillpipe. They provide bit weight and stability for faster drilling rates and aid in drilling a smooth, straight or smooth, curved hole. Stabilizers give varying degrees of rigidity or limberness. Heavy BHAs are a concentrated weight at the bottom of the drillstring, so the hole drills vertically downward aided by gravity. Directional equipment on the bottom part of assemblies causes the bit to drill directionally away from axis of the immediate upperhole section.

DESIGN AND CRITERIA

Design criteria are general guidelines based on equipment specifications and operating experience for building the bottomhole assembly (BHA). BHAs should be designed for maximum efficiency. Assembly efficiency is a measure of how well the assembly does its design function during drilling. This depends on operation, deviation, and stabilization tools, as well as formation dip, hardness, and drillability. Computer programs can aid in the design process.

BHAs can be exposed to extremely harsh operating conditions, depending upon the angle and number of bends and turns, depth, and related factors. They have a larger diameter and are stronger than the drillpipe string, so tensile strength usually is not a factor.

The simplest BHA, the limber assembly, is a string of drill collars with a bit on bottom. Larger, full-sized drill collars should be placed in the lower part of the assembly, and worn, smaller collars should be located in the upper part. Stabilizers and other equipment should be connected in various combinations to the drill collars for building different assemblies. The diverting equipment should be placed in the lower section of the assembly, where it has the most influence on directional control. Small variations in tool spacing may have a large effect on BHA efficiency. Pony drill collars and spacer subs are used for correct spacing. The amount of stabilization and resulting assembly rigidity should be minimized as much as possible without sacrificing efficiency. Connections are points of weakness and potential failure. Crossover subs and other connections should be eliminated whenever possible. Additional equipment such as keyseat wipers and drilling jar-bumpers should be installed in the upper part of the assembly.

Deflection tools such as a bent sub or bent-housing motor change the direction of the hole. These are identified by degrees of bend, the actual angle built into the tool. This is a reference and not the actual angular rate of change made during drilling. Normally the actual angular change is considerably higher than the reference bend angle. The amount of change depends upon tool placement in the drilling assembly, equipment used, formations, and operating parameters. For example, a 2° bent sub will curve the hole about 3°–6°/100 ft drilled, depending upon assembly design and the other factors noted. Deflection tools in combination cause higher rates of change. A 1° bent sub above a 1° bent-housing motor in a standard assembly will curve the hole about 8°/100 ft. The same assembly with a 2° bent sub and 3° bent-housing motor will curve the hole about 20°/100 ft. Supporting the bent-housing motor increases the build rate to about 24°/100 ft.

The number of nonmagnetic drill collars, usually one to three, depends upon hole and assembly diameters, drift, direction, and the earth's magnet lines of declination at the drill site. It is important to ensure that the steel in the remainder of the BHA does not affect compass-type course measurements. The actual number required may be found from empirical charts and tables. Nonmagnetic stabilizers should be used if necessary. These stabilizers may have some magnetic material such as the hard facing. The measuring instrument receptacle should be placed so that the compass is near, or slightly below, the center of the nonmagnetic collar section and as near to the bottom of the assembly as possible. The position may vary depending upon the equipment on the bottomhole assembly. It should be noted that drill tools develop or gain magnetism due to movement in the hole.

There is a high incidence of keyseating and wall sticking in directional and horizontal drilling compared to vertical drilling. Spiral or fluted collars should be used when applicable. Torque and drag normally are higher in directional wells and are main considerations in drillstring design. Torque and drag in common directional holes are about 15–30% more than that of a vertical hole at the same equivalent depth (TVD). Higher values are not uncommon on more complicated directional designs and on most horizontal holes. Assembly weight should be reduced to minimize high drag and torque. Part of the regular collars can be replaced with heavy drillpipe, especially for drilling in horizontal holes. Additional bit weight may be obtained by concentrating heavy BHA components near the bit. Part of the collars in the upper part of the assembly can be replaced with heavy drillpipe. One acceptable practice is to place heavy drillpipe above drilling jars. Heavy drillpipe generally should not be used in the bottom part of most BHAs because it is less rigid.

Some BHAs (split assemblies) can be divided into two parts for severe conditions, moving the upper part of the BHA to the vertical hole section in some directional and many horizontal holes. These can be connected together with compression pipe, or sometimes heavy or regular drillpipe. This is highly effective but should be used with caution as described in Chapter 5.

Wall sticking in the upper drill collar assembly should be prevented with heavy drillpipe or spiral drill collars, NOT STABILIZERS. It is important to remember that there is a higher risk of a drillstring failure in deviated holes; BHAs should be designed and operated carefully following API Bulletin 7G, *Drill Stem Design And Operating Limits* in most cases.

CLASSIFICATIONS

Bottomhole assemblies are subdivided into rotary and motor classifications (see Fig. 2–10). Rotary bottomhole assemblies are turned with a rotary or top drive. The bit on motor assemblies is turned with a turbine or positive displacement motor in the lower drill collar assembly while the drillstring remains stationary.

Bottomhole assemblies can be further divided into categories or types, some common to both classifications. The kind of equipment

Figure 2–10
Bottomhole assemblies

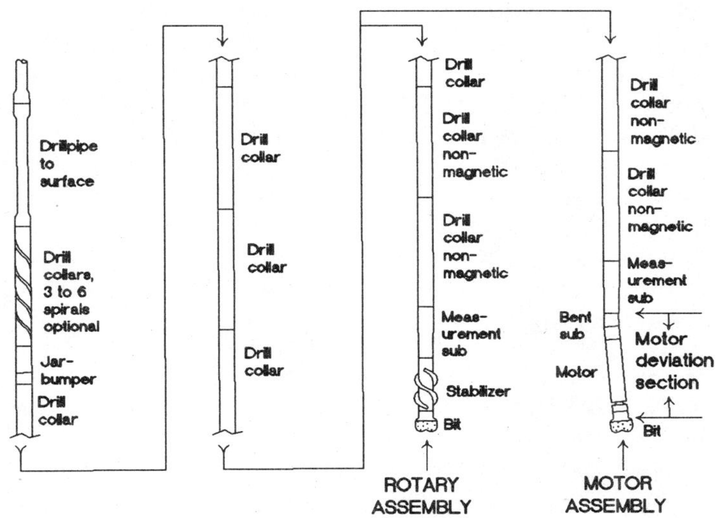

and position in the BHA normally decide the assembly type. For example, the various types of motor deviation sections in Figure 2–9 can replace the motor assembly portrayed in Figure 2–10. Conventionally, BHAs are named based on usage as listed in Table 2–3.

Table 2–3
Drilling Assembly Classifications.

Name (Usage)	Type
Limber	Rotary, motor less common
Deviation or sidetrack	Motor
Angle build	Rotary or motor
Angle drop	Rotary or motor
Angle hold or stiff	Rotary or motor
Reaming	Rotary
Coring	Rotary, motor less common
Open hole testing	Rotary
Fishing	Rotary

Limber, coring, and reaming assemblies are standard vertical drilling assemblies, but they are used in directional and horizontal drilling in special situations. A logical question is, Why have so many different assemblies if the deviation motor assembly drills in any direction? The answer is that another type of assembly often is more efficient for certain directional drilling operations, reducing costs and sometimes risk. For example, a deviation motor assembly can drop angle. But, an angle-drop rotary assembly often will drop angle faster, more economically, and with less risk. Another good example is the angle-hold rotary assembly that is very efficient in applicable conditions.

LIMBER

Limber rotary assemblies have standard drill collars, usually 12 to 20, connected together with a bit on bottom. First it is necessary to determine the exact number of drill collars needed based on collar weight, projected bit weight, and the location of the free point as described in the section concerning free point. A jar-bumper sub and keyseat wiper are commonly used as safety features. Limber assemblies are general-purpose assemblies commonly used in vertical drilling. They may be used in directional and horizontal operations for cleaning out the hole and when drilling without directional control. There is less risk of failure and sticking, and recovery by fishing is more successful. Limber assemblies also serve as a base for constructing other assemblies, either by adding tools between the drill collars or to the bottom of the assembly.

Limber motor assemblies are similar to limber rotary assemblies except that a turbine or positive displacement motor is positioned immediately above the bit. They serve similar purposes and may increase the penetration rate in vertical and some directional drilling.

Bit rotation is a major factor affecting penetration rate. A rotary or top drive rotates the bit about 50 to 150 revolutions per minute (rpm). A positive displacement motor turns the bit about 250 to 450 rpm, and the turbine will turn it faster. The normal practice is to turn the assembly slowly with the rotary or top drive when drilling straight, vertical or straight, inclined sections with the motor. The high rpm of the motor gives the main drilling action and may increase appreciably the amount of hole drilled. The basic criteria for selecting a motor assembly versus a rotary assembly depend on the incremental amount of hole drilled and additional cost of the turbine or motor.

DEVIATION

Deviation motor assemblies change the direction of the hole, drilling the new hole in a different heading. They deviate, sidetrack, and correct hole direction as described previously. Angle-build rates are 2°–5°/100 ft for regular assemblies and more for high angle-build assemblies.

ANGLE BUILD

Angle-build assemblies build or increase the angle of the hole in the vertical direction as previously described. Regular angle-build rotary assemblies build angle at 2°–5°/100 ft in a wellbore with an established buildup curvature. The build rate may be adjusted by changing the position of the stabilizer near the bit. Maximum efficiency is obtained in holes with inclinations of 10°–25°.

The angle-build motor assembly has a motor or turbine immediately above the bit. The most common assembly has a bent sub above the motor. There are many variations for building angle at different rates such as pads under bend sections, fixed stabilizer blades under the lower motor section, bent housings, and combination tools. Buildup angles range from a few degrees to more than 20°/100 ft for more aggressive assemblies. Drainhole angle-build assemblies are a special type guided by a whipstock mechanism.

The Hooligan angle-build assembly is a special angle-building assembly. It is similar to the standard angle-build assembly except that it has a short section, 30–50 ft, of smaller diameter pipe or collars above the near-bit stabilizer. The smaller diameter pipe is flexible and bends more easily to increase the angle-build rate. The motor version has a positive displacement motor immediately above the bit. Hooligan assemblies build angle at 3°–8°/100 ft. They are structurally weaker than the other assemblies because of the smaller diameter pipe. They must be operated carefully to prevent failure. The rate of angle build may be increased with higher bit weight and lower rotary speed.

ANGLE DROP

Angle-drop assemblies, often called pendulums, reduce the angle of the hole in the vertical plane. They are also used in vertical drilling to drill vertically downward where there is a formation tendency to cause the hole to deviate. A regular pendulum has one stabilizer located about 60 ft above the bit as previously described in the directional control section. These assemblies drop angle at 2°–4°/100 ft in high-angle holes and at a lesser rate as the angle of the hole decreases. The drop rate reduces at lower hole angles. The drop rate can be changed by adjusting the distance between the stabilizer and bit. A bit with aggressive side-cutting action drills more on the low side of the hole, increasing the angle-drop rate. Angle-drop rotary assemblies are very efficient. Angle-drop motor assemblies are similar to rotary assemblies but have a turbine or motor placed above the bit.

A packed-hole pendulum is similar to a regular pendulum except that it has two stabilizers, normally next to each other or separated by a pony drill collar. The extra stabilizer gives additional support at the fulcrum point for out-of-gauge holes or when the single stabilizer embeds in soft formations. Another variation has an extra stabilizer placed 30 ft above the fulcrum stabilizer. This reduces the drill collar sag above the fulcrum stabilizer and increases the aggressiveness of the assembly.

The forced pendulum is similar to the regular pendulum except that the stabilizer is closer to the bit, usually within 30–45 ft. Additional weight flexes the collars so that they bend, causing the bit to drill a downward curve. The forced pendulum is used instead of the regular pendulum to increase the drilling rate and still permit angle-dropping when conditions are favorable. The stabilizer spacing, bit weight, and rotary speed may be adjusted to improve performance.

REAMING

Reaming assemblies straighten and smooth crooked holes, restore undergauge holes to gauge, smooth out irregularities in the wellbores, and remove keyseats. They reduce excessive hole curvature over short intervals such as those entering and exiting a sharply curved section. Most reaming is a high-risk operation as described in the section on reaming. Two basic types of reaming assemblies are drill-collar reaming assemblies and string reamer assemblies.

The drill-collar reaming assembly has a near-bit reamer and reamers on top of the first and second drill collars above the bit. More aggressive reaming assemblies are run for severe hole conditions. They may have a near-bit reamer, a pony collar located above the bit with another reamer above it, and reamers on top of the next several drill collars. It is always a good practice to use a pilot-type hole opener on bottom instead of a bit for severe reaming conditions. This reduces the risk of accidentally sidetracking.

String reamers include one or more reamers positioned in the drillpipe string above the BHA. They rotate with the drillstring for reaming with the bit off bottom. They can be positioned in the drillpipe string so that they ream through specific hole sections while continuing drilling with the bit on bottom.

HOLD

Hold or stiff rotary assemblies maintain the drift and direction of the wellbore while drilling vertical or inclined hole sections. Hold motor assemblies are similar except that a turbine or motor on bottom rotates the bit. One modification is a slight build or steerable assembly. It is similar to angle-build motor assemblies but is designed for a low angle of build. The construction and action of hold assemblies was described in the directional control section.

MISCELLANEOUS

Coring assemblies cut and retrieve core samples of the formation. A core rotary assembly has a core barrel connected to the bottom of a shortened limber drilling assembly. A coring motor assembly is similar but has a motor connected above the core barrel (see Fig. 2–11). Open-hole formation-testing assemblies test the formation with testing tools connected to the bottom of a limber rotary assembly. Fishing assemblies use a limber rotary assembly with fishing tools connected to the bottom. A jar-bumper sub must be used on most assemblies.

HORIZONTAL APPLICATIONS

High-angle and horizontal hole assemblies have the same operating principles as those previously described. The assemblies generally are more complex and operate under demanding condi-

Figure 2–11
Coring tools
(courtesy of Eastman Christensen, a Baker-Hughes company)

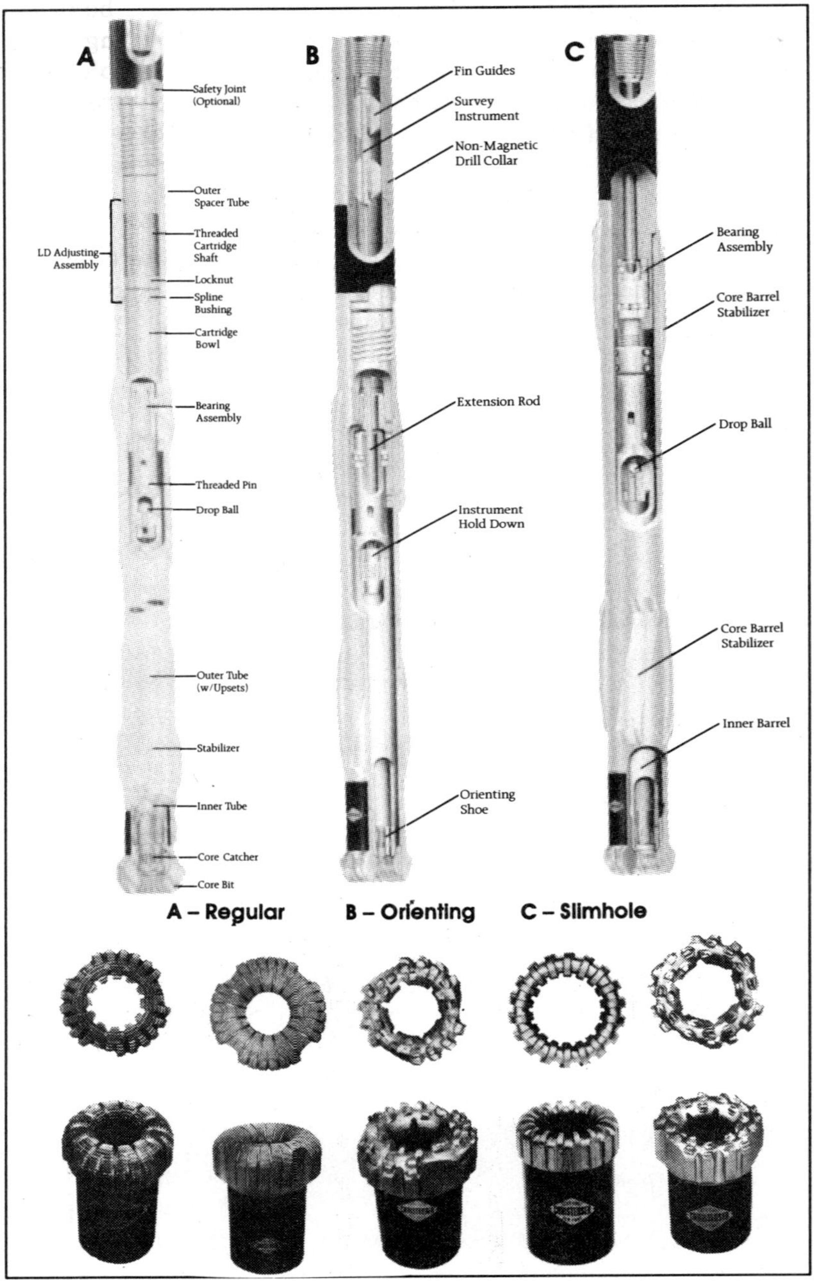

tions so predictability, reliability, and good motor performance are critical. Most are motor assemblies.

More common bottomhole assemblies (BHAs) have one point of leverage contact with the wall of the hole, such as a bent sub, excluding the bit. Directional control is fair at low angles but becomes much harder at higher angles. BHAs with multiple points of wall contact above the bit can be controlled more accurately, even at higher angles (greater than 50°; see Fig. 2–12). Three points of contact define a constant arc of curvature corresponding to the desired build rate of the motor assembly. These assemblies build angle predictably, even at high rates.

Motor assemblies can be fixed or adjustable. These refer to fulcrum extensions of fixed or adjustable thicknesses. Fixed assemblies have two aligned fulcrum supports for building angle accurately and at higher rates than adjustable assemblies. Adjustable

Figure 2–12
Motor deviation assembly turn radius

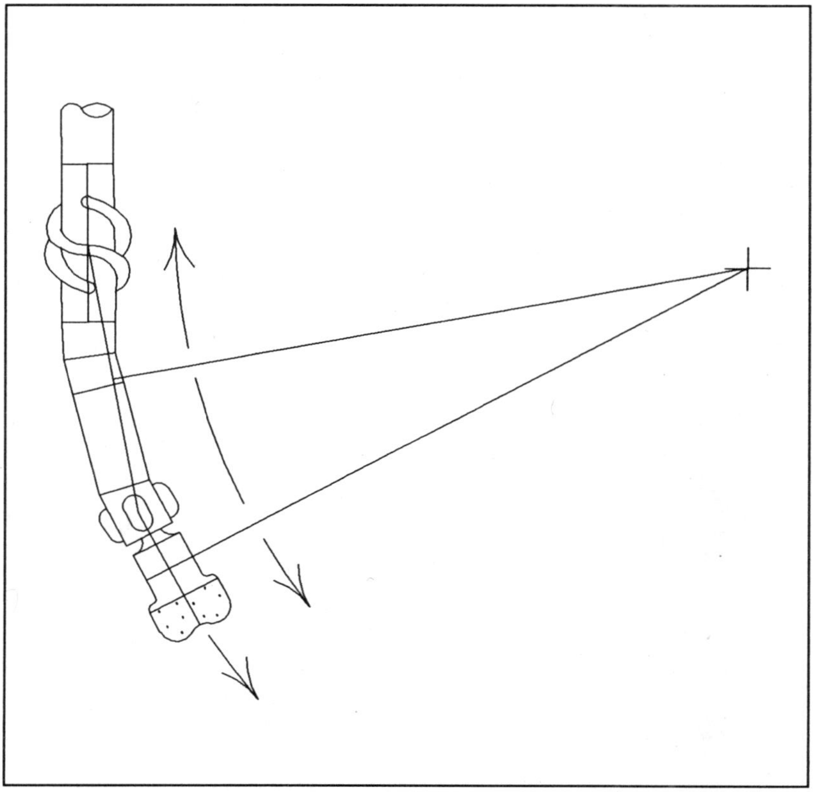

assemblies are more flexible for use in various situations, especially the steerable versions.

The term "steerable" has a special meaning in the oil and gas industry. Most motor assemblies are steerable in the sense that turning the drillstring changes the course in order to drill the hole in the desired direction. Steerable assemblies as used in the industry drill in one of two modes. The bit may be rotated with a motor or turbine while preventing the drillstring from rotating, so the drillstring slides downward while drilling the hole. The bit may be guided by turning the drillstring as required for controlling direction while drilling in this sliding mode, sometimes called oriented drilling.

With the second method, steerable mode, the entire drillstring rotates similarly to regular drilling. The bit turns by the motor and drillstring rotation. This action drills the hole straight ahead, not necessarily vertically, subject to formation effects. The main drilling action is by the motor rotating at hundreds of revolutions per minute (rpm) versus 5–15 rpm for the drillstring. Bit offset increases the hole diameter a small amount while rotating the drillstring. This normally is not a problem. Thus steerable assemblies provide for control of the direction of the hole, or allow faster drilling when the hole direction is satisfactory, both without making a trip to change the BHA.

Steerable assemblies have a low bit offset, the perpendicular distance in the horizontal direction from the centerline of the drillstring to the centerline of the bit (see Fig. 2–13). Most conventional deviation assemblies with a bent sub cannot be rotated because of high bit offset. The bent sub positioned above a standard positive displacement motor is a longer distance from the bit. This increases bit offset and exposes the lower part of the assembly to higher stresses when rotating the entire assembly. The net result is a shorter tool life, higher risk of a failure, and a possible fishing situation.

Assemblies with double-jointed motors can be rotated. Some motor assemblies that build up to 10°/100 ft may be steerable dependent upon equipment and spacing. Steerable assemblies can be highly effective, especially for drilling the curved and horizontal hole sections. Tangent sections described in Chapter 1 may be eliminated.

MEASUREMENT INSTRUMENTS

Measuring instruments record drift, direction, and tool face, the basic measurements for directional and horizontal operations.

Figure 2–13
Bit offset

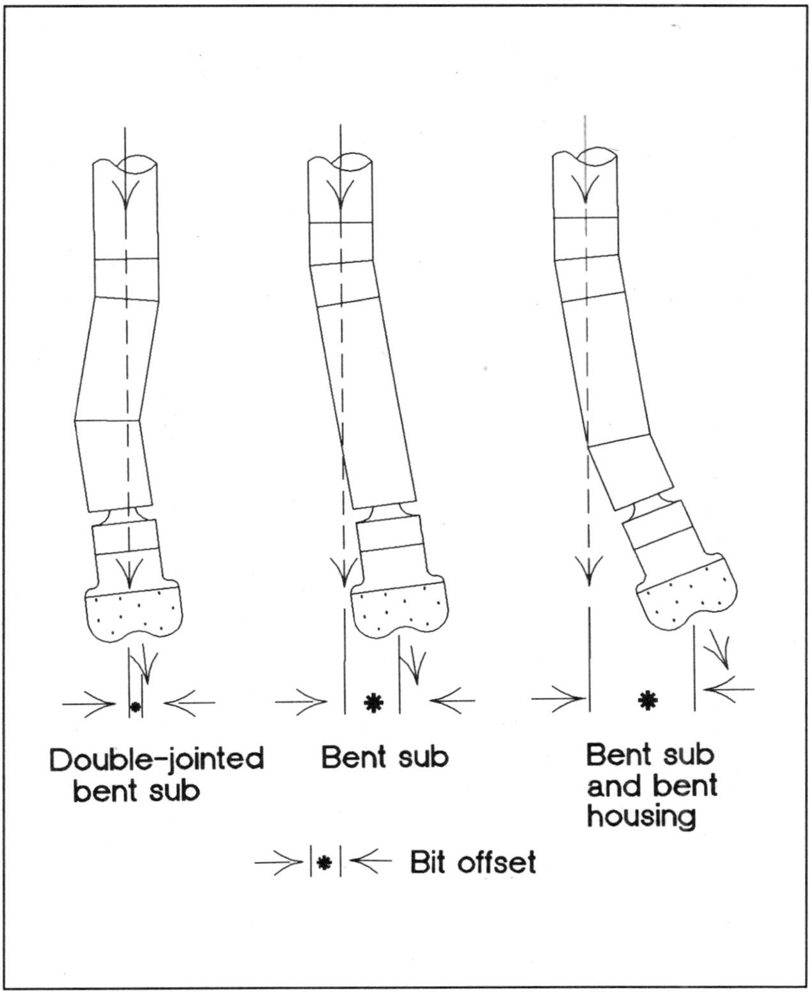

Measurement while drilling (MWD) instruments also record other data. Regular surveying instruments operate in temperature environments up to 250°–300° F. Higher temperatures degrade the electronics, batteries, and other parts of the measuring equipment. Heat shields insulate instruments and allow operation at higher well temperatures for a limited time.

DRIFT INDICATOR

The drift indicator was the first reliable instrument to measure the drift or angle of inclination of the wellbore. It does not record direction. A modified version of the tool is in use today. It has a free hanging plumb bob with a pin on the bottom. This suspends over a paper disk marked with concentric circles calibrated in degrees. A timing device actuates a mechanism that causes a pin to puncture the disk.

There are various modifications. One has a light source and light sensitive disk. Another records two measurements. After recording the first measurement, the disk rotates a half turn and records a second measurement as a verification of the first measurement. A motion sensor replaces the timer on drift indicators. It senses motion and will not actuate until the measuring instrument is at rest (motionless) for a predetermined period of time, usually about 30 seconds. This system has the advantages of fewer recording failures, less surveying time, and reduced risk of sticking.

In operation, the timer is set as required to allow time for running, and positioning with an interval so that the plumb bob can come to a complete rest. The instrument is placed in a centralized position inside a steel container. The carrier is lowered into the cased or open hole on a single-strand wireline to the measurement depth, where it aligns with the axes of the hole. The drift angle is recorded, the tool is lifted out of the hole and the disk is examined. The position of the point on the disk chart is the drift angle of the wellbore at the measurement point. Drift may be recorded at other points by repeating the procedure. In another method, the instrument is dropped inside the drillpipe before tripping and recovered after pulling the pipe. This instrument is commonly used for regular vertical drilling and in some common directional operations.

MAGNETIC SINGLE-SHOT

The magnetic single-shot measures both the drift and compass direction of the wellbore (see Fig. 2–14). The instrument has a precision floating compass, a device to superimpose concentric circles calibrated in degrees with a plumb-bob-type indicator. A camera photographs the plumb bob and compass face to record both drift and direction. Otherwise the magnetic single-shot is similar to the drift indicator and operates similarly. It cannot record compass directions inside steel pipe or casing because they blank off the earth's magnetic lines of force. It records measurements in the open hole or inside nonmagnetic drill collars. It was first used

Figure 2–14
Magnetic single-shot
(courtesy of Eastman Christensen, a Baker-Hughes company)

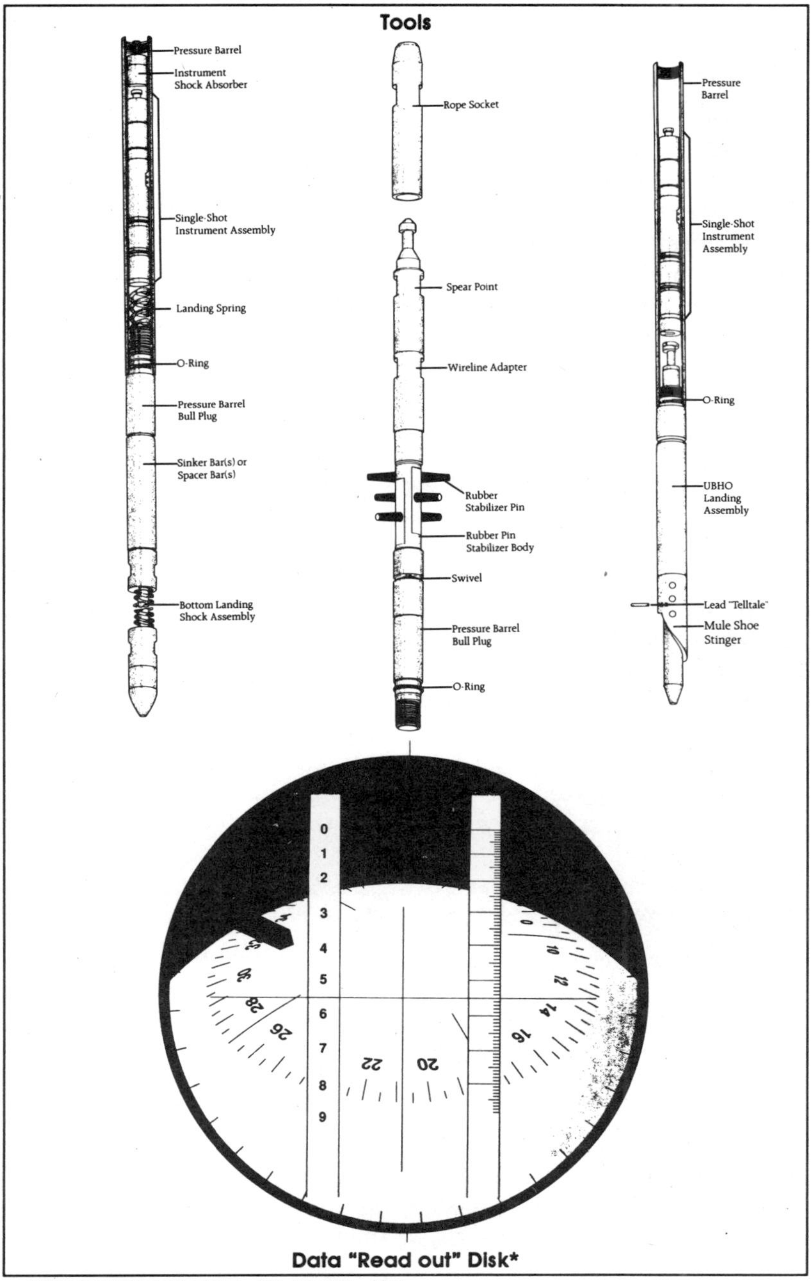

* Disk reading is inclinations = 55°, Azimuth = 208°.

in an old method of orienting by the "high side." A special version of this tool has limited use for high- or low-side orientation as described in Chapter 3.

A later version of the magnetic single-shot includes a pointer that indicates the direction of the tool face. It is in a fixed direction relative to the measuring instrument. The measuring instrument fits in a fixed, specific position inside the carrier container. The carrier has a muleshoe guide on bottom. When the carrier is lowered into the hole, this guide fits over a key slot in the orientation or measurement sub connected to the deviation assembly. This aligns the pointer relative to the key slot. Either the key slot should be aligned with the tool face when connecting the measurement sub in the deviation assembly, or the relative difference should be measured. During operation, a measurement records drift and magnetic direction of the wellbore and the relative direction of the tool face.

Sometimes the muleshoe and keylock system restrict flow rates. An improved version replaces the muleshoe and keylock with a magnetic tool face indicating pointer. The measuring instrument has two compasses. One is the floating type for drift and direction. The other is a needle-type tool face pointer. The measurement or orientation sub has two rows of small magnets positioned vertically along the axis of the sub and 180° apart. The magnets in each row are a few inches apart. Magnets in one row have their north pole facing outward from the center of the sub. Magnets in the other row have their south pole facing outward. This creates an induced magnetic field for the magnetic tool face indicating pointer. Then the angular difference between the tool face and the rows of magnets is measured. As with the muleshoe version, drift and magnetic direction of the wellbore and the relative direction of the tool face are recorded.

Magnetic single-shots are designed to measure angles within specific ranges. For example a 5° instrument measures drift angles between 0° and 5°. Likewise a 30° instrument measures angles between 0° and 30°. Generally an instrument is selected so that the drift angle is in the upper one half of the range for the best reading accuracy. Various instruments have different displays. The plumb bob position usually is located by either a small *X* or cross hairs, or, less frequently, a dot enclosed in a small circle. Most instruments use the set of concentric circles for measuring drift. A small circle in the margin or a radial line (or both) indicate the direction of the tool face. Chapter 3 contains a description of the operations of various measuring instruments used for orientation.

MAGNETIC MULTISHOT

Magnetic multishot instruments record multiple measurements of the drift and direction of the wellbore on a single run into the hole (see Fig. 2–15). The instrument is basically a modified magnetic single-shot instrument with the single-frame camera replaced by a

Figure 2–15
Magnetic multishot
(courtesy of Eastman Christensen, a Baker-Hughes company)

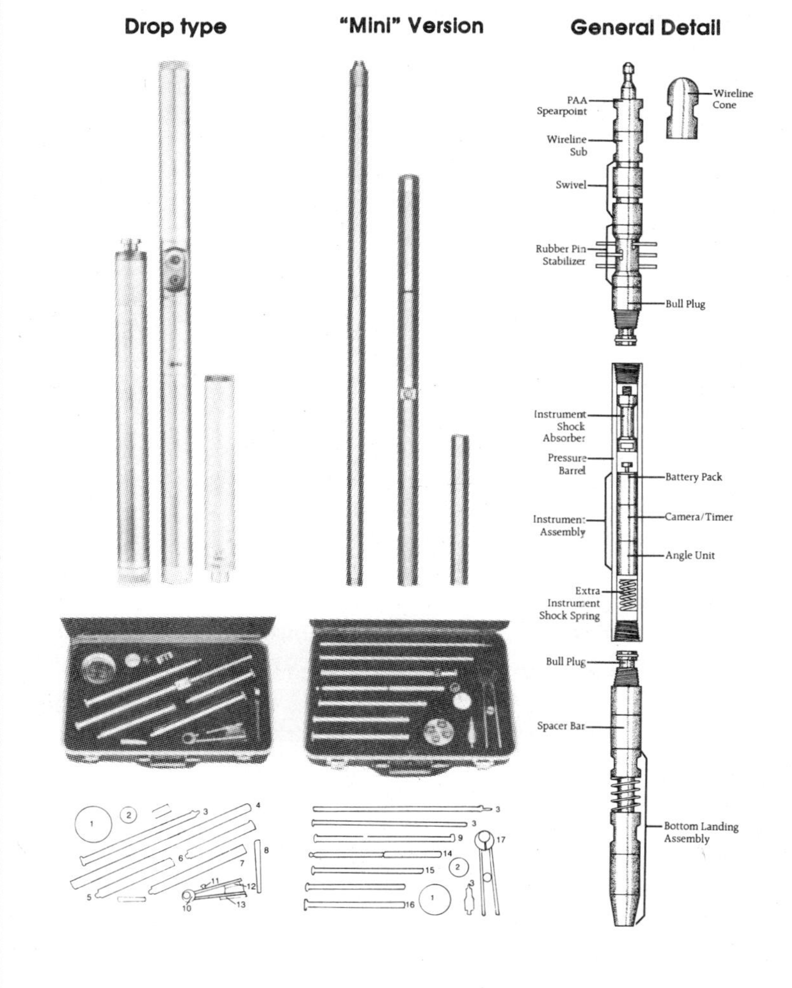

multiple-frame camera. There are various timing devices including the commonly used motion sensor.

The magnetic multishot operates similarly to single-shot instruments except that it records measurements in multiple-depth positions on one pass into the open or cased hole or inside the drillstring. It always measures drift angles, but must be in the open hole or inside nonmagnetic drill collars to measure compass directions. It is a common tool for wellbore surveys (see the wellbore survey section at the end of this chapter).

GYROSCOPE

Gyroscope instruments measure compass directions without using the earth's natural lines of magnetic force. Therefore, they can record compass directions inside steel drillpipe, drill collars, casing, or in the open hole. They also measure drift angles with regular or modified drift recording instruments. A spinning gyroscope maintains a base or fixed reference direction. During subsequent movements it measures relative changes of direction. The gyroscope operates similarly to other measuring instruments, the main difference being the method of measuring changes of drift.

The spin axis of the gyroscope is oriented to a fixed reference point in one method of operation. The tool is lowered into the open or cased hole or inside the drillstring, pausing for periodic measurements using a timer or the more common motion sensor. The tool is then pulled out of the hole and the measurements are read. These gyroscopes have errors such as gyro drift. An improved gyroscope development determines compass directions by relationships with the earth's rotation and force of gravity (see Fig. 2–16). It operates similarly to the earlier instrument, except it does not require orienting. Common uses are orientation and wellbore surveys in cased holes.

STEERING TOOL

Steering tools are common tools for recording measurements of drift, direction, and tool face during semicontinuous drilling. An instrument package contains a modified magnetic single-shot and other instruments. A coder converts data measurements to electrical pulses, and a sender transmits these to the surface through a shielded electric conduit. Surface equipment includes a decoder to convert the electrical pulses, digital or TV type displays, and

Figure 2–16
Gyroscope
(courtesy of Eastman Christensen, a Baker-Hughes company)

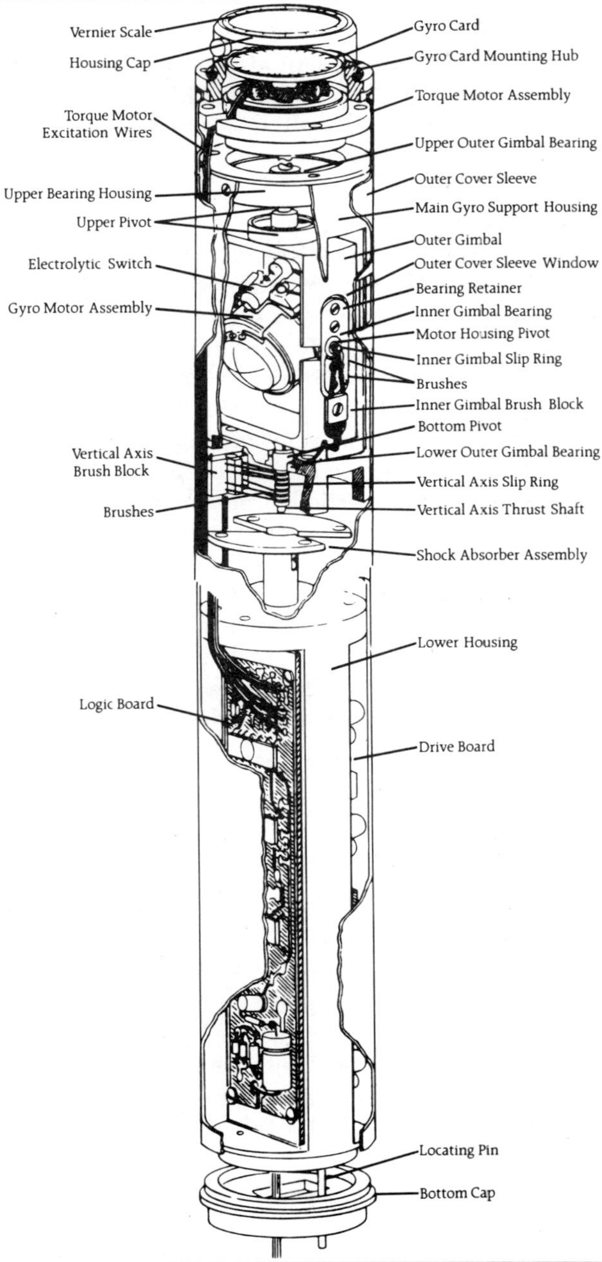

recorders. Thus measurements are available immediately at the surface for use to control hole direction. Steering tools eliminate many disadvantages of prior orientation systems, such as determining reactive torque.

In operation, the instrument package is lowered and raised with a shielded, electrical conduit (cable) on the reel of a winch positioned on a truck (cable truck). The package seats in a receiving or instrument sub in the deviating motor assembly and remains in the hole during drilling. Rotary assemblies cannot be used with steering tools. The steering tool cable is run in either a concentric or parallel configuration.

In the concentric configuration, first the drillstring with a directional motor assembly is lowered to (or near) the bottom of the hole. Then the instrument package is lowered through the drillstring with the cable and seated in the instrument sub. The annular space between the cable and drillpipe is sealed at the surface with a pressure pack off, so drilling fluid can be circulated down the drillstring for drilling. Then the directional or horizontal drilling operations proceed.

In the parallel configuration, the drillstring (with a directional motor assembly on bottom) is lowered partway into the hole. The instrument package is lowered inside the drillstring on a cable and seated in the measurement sub. The cable is passed through a side entry, or ported sub, out into the annular space. Then the cable and the drillstring are lowered together with the cable outside and parallel to the drillpipe string until the assembly is on or near bottom. Then directional or horizontal drilling operations proceed.

MEASUREMENT WHILE DRILLING

Measurement-while-drilling (MWD) records measurements at or near the bit while drilling continues (see Fig. 2–17). Data are transmitted immediately to the surface by pressure pulses in the mud column or by other methods that do not require an electrical conduit. MWD is highly applicable to drilling complex and extended-reach patterns and is almost a fundamental tool for horizontal drilling. It eliminates many problems common to other measurement systems.

A measuring instrument sub contains the MWD equipment and connects as part of the bottomhole assembly. Drift, direction, and tool face measuring instruments are modifications of standard

Figure 2–17
Measurement while drilling (MWD)
(courtesy of Halliburton)

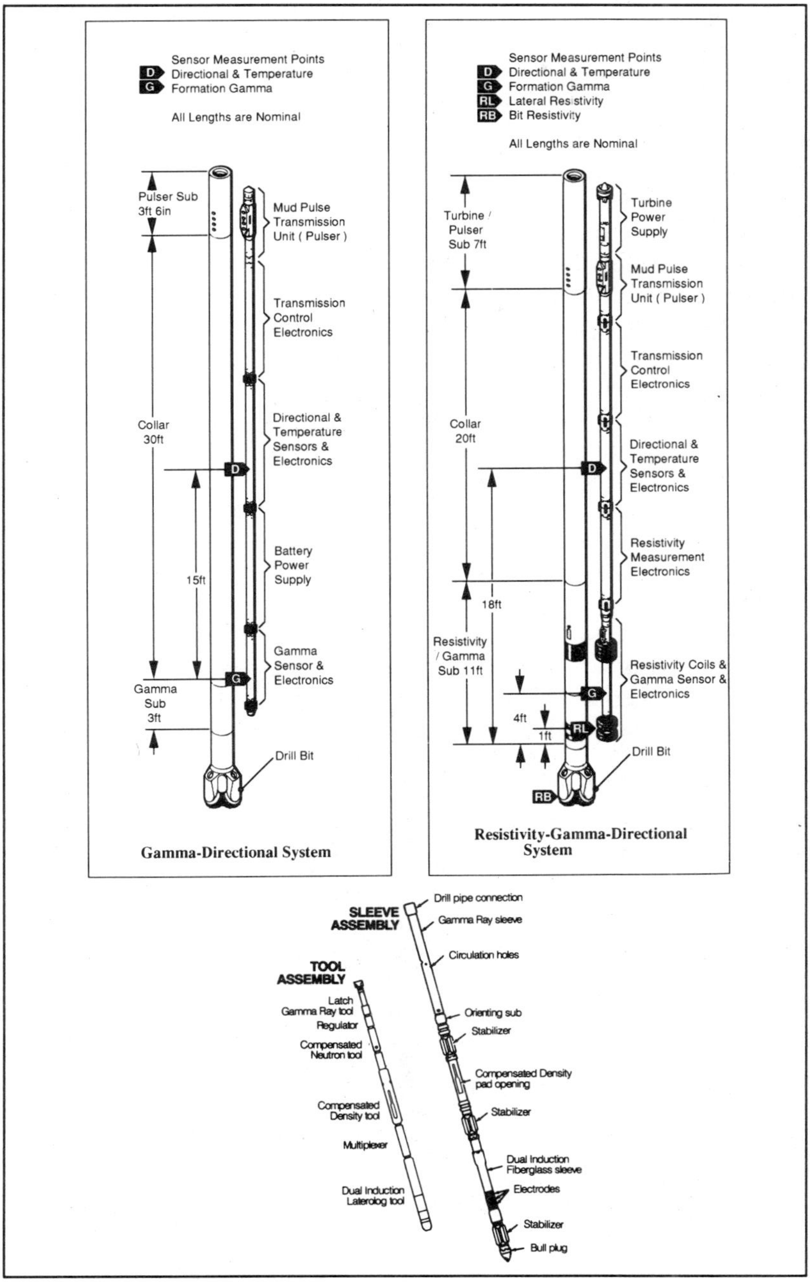

Gamma-Directional System

Resistivity-Gamma-Directional System

tools described earlier. Some systems provide for measuring other data with additional sensors. During drilling, measurements and other data are recorded, converted by a coder, stored in a storage device, and transmitted to the surface. The instrument package has batteries or a small turbine generator driven by the circulating mud for a power supply.

A common system has a mud pulser that receives stored data and converts it to high-frequency pressure pulses in the mud column, using mud pressure differentials between the inside and

Figure 2–18
Pulser unit
(courtesy of Halliburton)

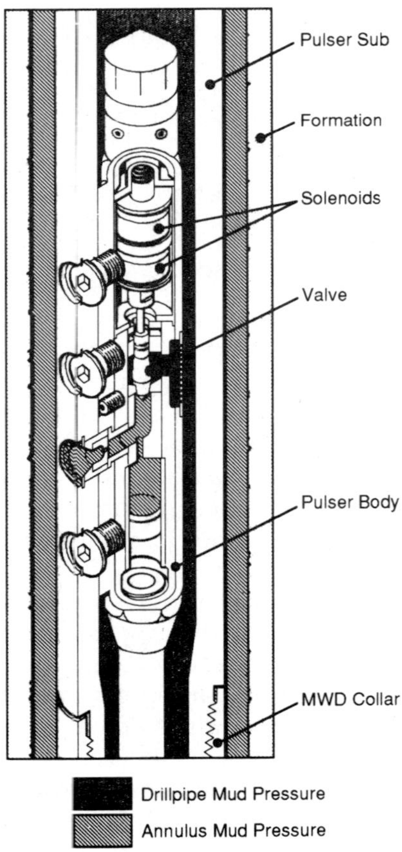

outside of the the drill collars (see Fig. 2–18). Pressure pulses travel through the mud column to a sensitive pressure detector at the surface. Surface equipment includes a decoder to convert the pressure pulses to electrical pulses and digital or TV-type displays and recorders. Drift, direction, and tool face measurements are immediately available for guidance to control the hole direction.

Advanced MWD systems measure and transmit a variety of data. Tools from different equipment suppliers have sensors on the BHA for measuring one or more of the different data as summarized in Table 2–4.

Table 2–4
Logging While Drilling (LWD).

Sensor	Data Produced
Rate of penetration	Drilling rate, ft/min or ft/hr
Rotary or bit speed	Revolutions per minute (rpm)
Mechanical efficiency log	Monitors bit condition
Sticking pipe indicator	Monitors friction losses
Strain gauge	Weight on bit, torque, bending moment
Temperature	Bottomhole mud temperature
Pressure	Bottomhole hydrostatic mud pressure
Gamma ray	Lithologic log
Resistivity	Short normal, focused resistivity
Conductivity	Induction, high-frequency conductivity
Neutron	Porosity log

The term MWD (measurement-while-drilling) refers to directional measurements, and LWD (logging-while-drilling) refers to recording other data. Some data recording systems require a water-base fluid. Other information may be calculated from the recorded data. This includes density, formation fracture gradient, formation fluids, formation fluid pressures, pore pressure plots, and fluid loss to the formation.

Measurement while drilling is the latest measurement tool and records measurements while drilling continuously. It is used in directional drilling and for most high-angle and horizontal drilling projects. It has significant advantages over earlier systems, especially for recording and transmitting data. Surveys require only a few minutes despite depth or inclination. Magnetic single-shots require at least 30 minutes at shallow depths plus the risk of hole problems. MWD eliminates most of the problems encountered in steering tool operations. MWD tool reliability and accuracy is an important feature and an increasing number of operators accept MWD accuracy. The formation logging feature can aid in determining the bottomhole location by recording geological markers (correlation points from other wells).

MWD equipment is available from several suppliers. There are various MWD systems with different measuring and data transmission systems. Some tools have limitations; for example, they cannot operate with a high content of lost circulation material in mud. Many operate slightly differently and have different advantages and limitations. It is important to be certain the tool selected fits the drilling conditions and measurement requirements.

Specifications of a combination MWD and LWD instrument system are included in Table 2–5.

Table 2–5
MWD-LWD Measuring Instrument Specifications.
(courtesy of Halliburton)

MEASURED PARAMETERS	RANGE OF MEASUREMENTS
Lateral resistivity	0.2–1,000 ohm/m
Bit resistivity	0.2–1,000 ohm/m
Gamma ray	0–500 AAPI
Focused gamma ray	0–500 AAPI
Neutron porosity	0 to 100 limestone porosity units
Formation density	1.0 to 3.0 gm/cm^3
Inclination	0–180°
Azimuth	0–360°
Tool face	0–360°
Internal temperature	0–200°C
Drillpipe temperature	0–200 °C
Annulus pressure	0–20,000 psi
Drillpipe pressure	0–20,000 psi
Pressure drop across bit and bHA	0–5,000 psi

TOOL FEATURES

Downhole memory capacity — 120 Kbytes
Memory data acquisition rates — Every 5, 10, 20, or 40 seconds
Mud pulse data transmission rates — 0.5, 1, or 2 bits per second
Mud pulse data transmission sequence — logging only, survey only, logging-survey, survey-steering, logging-survey-steering

DOWNHOLE TOOL PROGRAMMABILITY

Downhole memory
Activation/deactivation
Downhole memory data acquisition rate
Mud pulse data transmission rate
Mud pulse data transmission sequence

OPERATING PARAMETERS

Mud flow = 250 — 1,500 GPM
Mud pumps — Duplex or Triplex
Mud sand content — Maximum 1% (5% with low-volume flow)
Required bit pressure drop — 500 psi. Restrictor sub available at low flow rates to yield necessary pressure drop across the tool.
Maximum operating temperature— 150°C (170°–190°C)*
Maximum tool pressure — 20,000 psi

SURFACE EQUIPMENT

Standard field unit	8 ft x 12 ft x 6 in.; 16,000 lb. Computer system, printers, plotters and operator terminal.
Advanced field unit	10 ft x 28 ft x 9 ft; 30,800 lb), Wellsite data collator, multi-user database, data transmission and high-resolution graphics.
Rig floor display unit	Digital display for azimuth, inclina tion and tool-face orientation.

TUBULAR LENGTHS

Collar — 15 ft
Turbine/pulser sub — 7 ft
Mandrel — 11 ft
Complete tool (with crossover) — 38 ft

*Formation Temperature when Circulating

INSTRUMENT ACCURACY

Instrument accuracy is a fundamental necessity for directional and horizontal drilling. Errors may be very significant at times and less important at other times. For example, drilling kill wells may require very accurate measurements. Accuracy can cause substantial differences in the mapping and interpretation of small, deep structures, especially in faulted areas. Accuracy is also very important in mapping and determining reserves. Small errors over long distances can be substantial; it is always important to strive for optimum accuracy.

There are various sources of inaccuracies, ranging from the basic accuracy of the instrument, including calibration and usage errors, to interpretation of measurements. These have been the subject of many studies. Modern survey instruments are sufficiently accurate for most field purposes. Drift angle measurement errors range from a few feet (or less) per thousand feet of hole at low angles to 5 to 7 ft at higher angles. Direction measurement errors correspond to drift angle errors and may be slightly higher. Instrument accuracy decreases at higher angles, especially in horizontal wells.

Instrument accuracy can be tested in the wellbore. Surveys should be checked by repeating measurements with different brands and types of instruments. Another procedure selects several reference points at different depths, taking multiple, accurate measurements, usually with a single-shot instrument. The variance from the average should be reviewed. The same procedure is used with measurement while drilling equipment for determining relative accuracy. MWD logging instruments have been checked extensively against other types of logging. These and other measuring instruments normally are sufficiently accurate for most common directional projects.

WELLBORE SURVEYS

Wellbore surveys are measurements of the drift and direction of the wellbore. Vertical wells, usually drift surveys, are surveyed to control inclination and prevent the hole from becoming crooked. Drift and direction surveys are recorded in high-risk wells so that the wellbore can be located by a kill well if the well blows out. Drilling contracts require surveys in standard clauses. Gyroscopic surveys of existing cased wells can locate the kickoff point relative to the surface location. Existing cased wells should be surveyed if there is a dispute about the location of the bottom of the hole with respect to property lines and surface ownership. Most regulatory agencies require individual surveys during drilling or a later wellbore survey.

The more economical magnetic multishot is used to measure drift angle when omitting direction measurements. Gyroscope instruments are used to record direction measurements in drillpipe and cased wellbores. Surveys can be done with a wireline or drillstring, depending upon the type of survey and well status. The instruments are lowered on a wireline to total depth and then lifted, stopping periodically to record measurements. Alternately, the instruments can be placed in the bottom of the drillstring before tripping. A nonmagnetic drill collar in the BHA can be used for directional measurements unless using a gyroscopic instrument. The drillstring is pulled out of the hole, stopping to record measurements at various depths. Wellbores are surveyed at intervals of about 200–500 ft, depending upon the accuracy required and the reason for the survey.

BIBLIOGRAPHY

American Petroleum Institute. *Recommended Practice for Drill Stem Design and Operating Limits* API RP7G. American Petroleum Institute. March 1973.

C. A. Bardin. "Remote-Controlled Bent Sub Aids Directional Drilling by Allowing Bend-Angle Change." *Oil & Gas Journal* (January 20, 1989): 76–80.

T. R. Bates Jr. and C. A. Martin. "Multisensor Measurements-While-Drilling Tool Improves Drilling Economics." *Oil & Gas Journal* (March 19, 1984): 119–138.

M. M. Clary and T. W. Stafford. *MWD Performance and Economic Benefits in the Zu Horizontal Drilling Program.* SPE/IADC 16171. Society of Petroleum Engineers, International Association of Drilling Contractors. New Orleans, LA, March 15–18, 1987.

K. T. Corbett and R. Dawson. "Drillstring Design for Directional Wells." *Oil & Gas Journal,* (April 30, 1984): 61–64.

F. V. DeLucia, *Benefits, Limitations, and Applicability of Steerable System Drilling.* SPE 18656. Society of Petroleum Engineers/International Association of Drilling Contractors. Drilling Conference. New Orleans, LA, March 1989.

R. Desbrandes. "Status Report: MWD Technology." *Petroleum Engineer International* 3 part series. (September–November 1988).

O. Dubrule and P. H. Nelson. "Evaluation of Directional Survey Errors at Prudhoe Bay." *SPE Drilling Engineering* (September 1987): 257–267.

J. H. Enenbach. "Directional Drilling Technology Strives for Speed and Accuracy." *Petroleum Engineer International* (September 1980): 124–132.

D. Eckstrom. "Bits with Diamond-Coated Inserts Reduce Gauge Problems." *Oil & Gas Journal* (June 17, 1991): 41–43.

R. Gross. "Apache Blowout Successfully Killed." *Drilling* (March 1984).

D. Gust. *Directional Drilling Evolving from Art to Science.* Calgary, Canada.

Halliburton Services. *Howco Cementing Tables.* Halliburton Services, A Halliburton Company, 1981.

J. E. Hansford and A. Lubinski. "Cumulative Fatigue Damage of Drillpipe in Dog Legs." *Transactions of the American Institute of Mining and Metallurgical Engineers* 237 (1966): I–359.

P. R. Hornbrook and C. M. Mason. "Improved Coiled-Tubing Squeeze-Cementing Techniques at Prudhoe Bay." *Journal Of Petroleum Technology* (April 1991): 455–458.

A. P. Jourdan, P. Armessen, and P. Rousselet. "Horizontal Well Operations — Part 4: Elf Has Set up Rules for Horizontal Drilling." *Oil & Gas Journal* (May 9, 1988): 33–40.

A. W. Kamp. *Downhole Telemetry from the User's Point of View.* SPE 11227. Society of Petroleum Engineers. 1982.

R. D. Kane. "High Alloy Tubulars Hold Promise for Sour Service Tolerance." *Petroleum Engineer International* (January 1983): 98–112.

E. Kral, et al. "Drillpipe Fracture." Part 1 and 2. *Oil & Gas Journal* (August 6 and 13, 1984).

A. Lubinski. "Factors Affecting the Angle of Inclination and Dog-Legging in Rotary Boreholes." *Oil & Gas Journal* (March 23, 1953).

T. W. McKay. *A Method for Designing a Complex Directional Drilling Program Applied in Cook Inlet, Alaska.* SPE 10056. Society of Petroleum Engineers. San Antonio, October 5–7, 1981.

K. Millheim. "Behavior of Multiple-Stabilizer Bottomhole Assemblies." Part 5. *Oil & Gas Journal* (January 1, 1979): 50-64.

K. Millheim. "Control Techniques for Medium-Soft and Medium Formations." Part 7. *Oil & Gas Journal* (January 29, 1979): 178–186.

K. K. Millheim, S. Jordan, and C. J. Ritter." Bottom Hole Assembly Analysis Using the Finite Element Method." *Journal of Petroleum Technology* (February 1978): 265–274.

T. J. Mitrou, et al. "Comparison of Magnetic Single-Shot Instruments with a Directional MWD System." *SPE Drilling Engineering* (April 1986): 163–168.

S. D. Moore. "The Coil Tubing Boom." *Petroleum Engineer International* (April 1991): 16–20.

G. Nazzi. "Horizontal Wells 2—Planning Matches Drilling Equipment to Objectives." *Oil & Gas Journal* (October 8, 1990): 110–118.

G. Pidcock and J. Daudey. "Gulf Canada Improves Drilling with MWD Techniques." *Petroleum Engineer International* (September 1988): 16–24.

A. A. Pogarskiy and A. M. Yasashin. "U.S.S.R. Turbodrilling ROP Exceeds U.S. Rate." *Oil & Gas Journal* (June 3, 1991).

A. C. Scott and B. E. MacDonald. *Determining Downhole Magnetic Interference on Directional Surveys.* SPE 7748. Society of Petroleum Engineers. Manama, Bahrain, March 25–29, 1979.

F. P. Shray. "LWD Detects Changes in Formation Parameters over Time." *Petroleum Engineer International* (April 1992): 24–31.

M. Stephenson. "Program Challenges Directional Survey Accuracy Claims." *Oil & Gas Journal* (August 20, 1984):112–120.

D. R. Tanguy and W. A. Zoeller. "Applications of Measurements While Drilling." SPE 10324. Society of Petroleum Engineers of the American Institute of Mining and Metallurgical Engineers. 1981.

B. A. Tarr and I.A. Graham. "North Sea Experience with Aluminum Drillpipe." *SPE Drilling Engineering* (December 1990): 317–324.

G. Tolle and T. Dellinger. "Mobil Identifies Extended-Reach-Drilling Advantages, Possibilities in North Sea." *Oil & Gas Journal* (May 26, 1986): 78–86.

P. Toutain. "Analyzing Drillstring Behavior." *World Oil* Part 1. (June 1981): 181–184; Part 2. (July 1981): 221–226; and Part 3. (September 1981): 143–145.

D. K. Trichel and M. P. Ohanian. *Unique Articulated Downhole Motor Holds Promising Future for Short Radius Horizontal Drilling.* SPE 20417. Society of Petroleum Engineers. New Orleans, LA, September 23–26, 1990.

E. Turner. "New Bit Designs Control Destructive Whirl Phenomenon." *World Oil* (June 1991): 60–65.

G. W. Uttecht and J. P. deWardt. "New Small-Diameter Inertial Gyroscopes Can Reduce Most Borehole Position Uncertainty." *Oil & Gas Journal* (April 11, 1983): 82–89.

J. E. Walstom, A. A. Brown, and R. P. Harvey. "An Analysis of Uncertainty in Directional Surveying." *Journal of Petroleum Technology* (December 1969).

C. J. M. Wolf and J. P. deWardt. "Borehole Position Uncertainty—Analysis of Measuring Methods and Derivation of Systematic Error Model." *Journal of Petroleum Technology* (December 1981).

H. B. Woods and A. Lubinski. "How to Determine Best Hole and Drill Collar Size." *Oil & Gas Journal* (June 7, 1954).

H. B. Woods. "Use of Stabilizers in Controlling Hole Deviation." *Drilling and Production Practices.* American Petroleum Institute. 1955.

J. Wright. *Rate Gyro Surveying of Wellbores in the Rocky Mountains.* SPE 11841. Society of Petroleum Engineers. Salt Lake City, UT, May 23–25, 1983.

W. A. Zoeller. *Pore Pressure Detection from MWD Gamma Ray.* SPE 12166. Society of Petroleum Engineers Annual Fall Meeting, San Francisco, CA, October 5–8, 1983.

CHAPTER 3

DEVIATION AND SIDETRACKING

SUMMARY

Deviating or sidetracking is the first step in most directional and horizontal drilling operations. Deviating is the procedure for starting at the bottom of an open or cased hole and drilling directionally. Sidetracking is similar, except that the new directionally drilled hole starts some distance from the bottom of the open or cased hole, sidetracking part of the original hole. Directional and sidetracking assemblies are oriented by first finding the direction and turn. Tool face correction, rotary torque, and bit walk must be allowed for when applicable.

The next step is to turn the assembly, pointing the tool face in the correct direction toward the target and begin to deviate or sidetrack. Magnetic single-shot, steering tool, or measurement while drilling instruments are used for measurements during orientation and later for directional and horizontal drilling. This is followed by deviating at the bottom of open and cased holes with a deviating assembly.

Sidetracking in open holes is accomplished by first plugging back with cement and then sidetracking with a sidetracking assembly. Some cased holes are sidetracked similarly after removing a section of casing by milling. Others may be sidetracked by cutting a hole through the side of the casing with a milling tool, using a whipstock as a guide. Slant holes start at the surface in an inclined direction pointed toward the target, drilling with a slant-hole rig. Other methods of deviating for specialized applications include curved or angled conductor or drive pipe, nudging, and by using small, oriented pilot holes.

SELECTING MEASUREMENT SYSTEMS

Three common orienting measuring systems are magnetic single-shot, steering tool, and measurement while drilling. Each system measures the compass direction and inclination or drift angle of the hole and direction of the tool face. Specific operations of the different measurement systems, with advantages and disadvantages, are included in the different deviation and sidetracking procedures described later in this chapter. Each has operational and other advantages and disadvantages. These should be evaluated in relation to the specific job and the most applicable system should be selected.

Magnetic single-shot is the oldest system in common use. The instrument has very good tool accuracy and reliability. It is less costly than other orientation systems. It also has disadvantages, such as being somewhat slow and its method of correcting for bit walk and reactive torque. The magnetic single-shot should be used in less difficult deviation, sidetracking, and for some correction runs, primarily for drilling directional patterns. Each survey takes from one to several hours depending upon depth. It may be necessary to repeat surveys due to miss runs or for verification.

There is less risk of failure and sticking while drilling with the magnetic single-shot system. Still, the drillstring must be motionless when recording measurements, so there is a risk of sticking. Risk increases in frequency and severity with increasing depth, while measuring in more complex patterns and when drilling problem formations. The drillstring should be moved a limited amount while running and retrieving the survey instrument except under special conditions. Deeper holes should be circulated simultaneously by using a pressure pack-off type circulating head. Good well control may be ensured by placing a full opening inside the blowout preventer on the top of the drillstring before running the measuring instruments in the hole. Reactive torque can be a problem as described in a later section.

The magnetic single-shot and other measurement systems, to some extent, have an inherent disadvantage. The measurement sub is about 10–25 ft above the bit depending upon the specific equipment and its position on the deviation assembly. The bit must be a safe distance of 5–15 ft above the bottom of the hole to reduce the risk of sticking while recording measurements. Therefore, measurements should be recorded at least 20–40 ft or higher off bottom. This requires drilling about 30–50 ft of directional hole before measurements detect the results of correction changes. This may cause problems in deviation and sidetracking, especially under conditions requiring close control. Otherwise, it is not a problem in regular directional drilling.

Steering tools record measurements of drift, direction, and tool face almost continuously while drilling and display them immediately on a surface monitor. Steering tool measuring instruments are used for drilling easier directional patterns. Concentric configuration should be limited to less difficult jobs. The steering tool is more costly, but it eliminates many disadvantages of the magnetic single-shot measurements, such as predicting the lead angle and compensating for reactive torque. Directional control is better and faster with more time spent drilling.

Measurements are not precisely accurate while drilling because of reactive torque and small assembly movements. They are sufficiently accurate for working. Accurate measurements should be obtained periodically for verification. Both the drilling and pumping should be suspended momentarily so that the downhole assembly comes to a complete rest for accurate measurements. Steering tools cost more than the magnetic single-shot, but increased efficiency may offset the higher cost. If there is a question about good well control, an inside blowout preventer should be used. Drillpipe rotation is limited due to a risk of pressure and mechanical sticking. Other disadvantages include using a cable truck, semicontinuous drilling, and those disadvantages related to the specific configuration.

The concentric configuration has a pack-off circulating head with pressure limitations that may cause extra cable wear, especially at elevated pressures. The instrument package can be changed without tripping if it fails. Drillpipe connections are tedious and time-consuming.

The parallel configuration requires a longer trip time, but it saves time making connections. The entire drillstring must be pulled to change the instrument package if it fails. There is higher risk of damaging the cable outside the drillpipe. It is preferable to run the exposed cable in a cased hole with drift angles less than about 60°. This allows the cable to be pulled out of the side-entry sub if the drillstring sticks. The side-entry sub may be a weak point in the pressure integrity of the drillstring. The parallel cable can either cause a fishing job or increase the severity of a fishing job if the drilling assembly sticks or the well kicks.

Measurement-while-drilling is the most advanced measurement system. It eliminates most of the problems of the other systems but costs more. Measurement-while-drilling is used for difficult deviation programs such as high-angle directional drilling and for most horizontal drilling. The data recording feature can be very advantageous.

ORIENTATION

Orientation is the combined procedure of selecting the correct direction and positioning the deviation assembly so that the bit points in that direction for drilling. It is a fundamental directional and horizontal drilling operation. Orientation normally refers to the horizontal direction when first deviating or sidetracking. Otherwise, it includes either horizontal or vertical directions or a combination of the two. A few holes are sidetracked without orientation, which is called blind sidetracking. The most common occurrence of this is bypassing a fish in either open or cased holes and sometimes sidetracking damaged casing. Modified orienting procedures are also used in coring.

Orientation is done when first deviating or sidetracking and repeated when the tool face changes to the wrong direction. Various conditions may cause the bit to drill in a different direction from the orientated direction. These include formation effects on hole direction, bit walk, reactive torque, and assembly performance and efficiency. Drilling procedures, especially bit weight and rotary speed, may change direction and drift. Sometimes the operator changes the target for various reasons, such as due to geological information revealed during drilling.

This section primarily covers orientation methods and finding the new direction of the tool face. The operations for changing the direction are included with the different deviation and sidetracking procedures described later in the chapter.

ORIENTATION METHODS

Three orientation methods are surface, indirect, and direct methods. The surface method was the first orienting procedure and is obsolete. It consisted of orienting the deviating assembly at the surface. Then the position was checked with a telescope and sighting device while lowering each joint or stand into the hole. Measurement accuracy was questionable and the procedure was tedious and time-consuming.

The indirect method of orientation uses direction changes relative to the high side or low side of the wellbore. It requires advance knowledge of the direction of the wellbore and resulting low and high sides. The high side of the hole is also the direction of the wellbore. (The plumb bob of the magnetic single-shot hangs to the low side of the hole and 180° opposite the direction of the wellbore.) Changes are measured from either the low or high side but must be consistent. This text describes the procedure referenced to the high side unless otherwise noted. The indirect method is seldom used except in a few cases for horizontal guidance while drilling high-angle and horizontal laterals with a stable drift. Indirect orienting procedures are described in a later section.

The first indirect tool had a mechanical device based upon a ring, key, and rolling ball for detecting and drilling on the low side. The tool, now obsolete, used a modified drift indicator. The next instrument, which still may be in limited use, was the regular magnetic single-shot with the muleshoe and without the tool face indicator. The latest measuring instrument is a modified magnetic single-shot. The floating-type compass seats opposite small orienting magnets in the instrument sub. Other measuring instruments can be modified and used.

The direct method is the most common and widely used procedure of orienting for directional and horizontal drilling. It is used in the remainder of this text unless otherwise noted. The direct method utilizes modern measuring instruments. Sometimes it is subdivided into the magnetic, gyroscopic, and steering tool methods. Still, measurements from these three measurement systems are basically similar. They record the drift and direction of the hole and the direction of the tool face. The main differences are their operation and means of recording and transmitting data.

The orienting procedure is simple in description and operations are straightforward. The deviating or sidetracking assembly is run into the hole near the bottom. The drift and direction of the hole and the direction of the tool face are measured. Then the drillstring is turned so that the tool face points to the correct direction. The tool face setting is verified with another measurement and deviating or directional drilling begins. The procedure is not complicated, especially for later measuring systems such as the steering tool and measurement-while-drilling. Corrections may be somewhat complicated with the magnetic single-shot but should not be a problem. Orientation should be conducted in a workmanlike manner. The main problems are in the operations as described for the various orienting procedures later in this chapter.

FINDING DIRECTION AND TURN

Finding the direction and amount of turn ranges from easy to complex, depending upon conditions. A build-and-turn guide serves to illustrate a few fundamentals (see Fig. 3–1). Note that the top of the chart is the high side, or direction of the wellbore and not north. The chart is only precisely accurate for a vertical hole. Accuracy decreases as the drift of the wellbore increases. The chart is sufficiently accurate for illustrative purposes at low drift angles of a few degrees.

Pointing the tool face in the vertical or upward direction will give the maximum build rate. Pointing the tool face to the right will give a maximum right turn. The tool face is pointed in the upper right quadrant for both building angle and turning to the right. If the tool face is pointed in the upper right quadrant and closer to the vertical, angle building increases with reduced right turn. Changing the tool face more to the right, within the same quadrant, decreases the angle-build rate and increases the right turn. The same reasoning applies to the other quadrants and points on the circle.

It must be remembered that points on the circle are referenced to the direction of the wellbore. For example, assume a wellbore direction of south, 30° west. The tool face is turned 45° to the right to south 75° west for building angle and turning to the right.

As noted, chart accuracy decreases as the drift angle increases. High drift angles are common, requiring a better method of prediction. This is accomplished by the use of vector diagrams. Vector analysis is beyond the scope of this book, but the procedure can be summarized briefly. A dogleg is calculated from the current wellbore drift and direction and force of the deviating tool. These are used to determine a change of direction and new drift angle at a deeper depth, based on turning the assembly a fixed amount. The "ouija board," similar to a special type of slide rule, was an early method for solving these. They can be solved graphically by vector diagrams, but the process is tedious and time-consuming. They are commonly solved with proprietary computer programs.

A major unknown is the effect of the formations. They affect the direction of the hole as covered in Chapter 4. The type of deflecting tools and the manner of operation also affect hole direction. Bit walk and reactive torque are additional factors. All of these must be considered when determining the direction for orientation.

REACTIVE TORQUE

Reactive torque is the counterreaction of the drillstring to torque caused by the bit and motor during drilling. This torque causes the

Figure 3–1
Build-and-turn guide

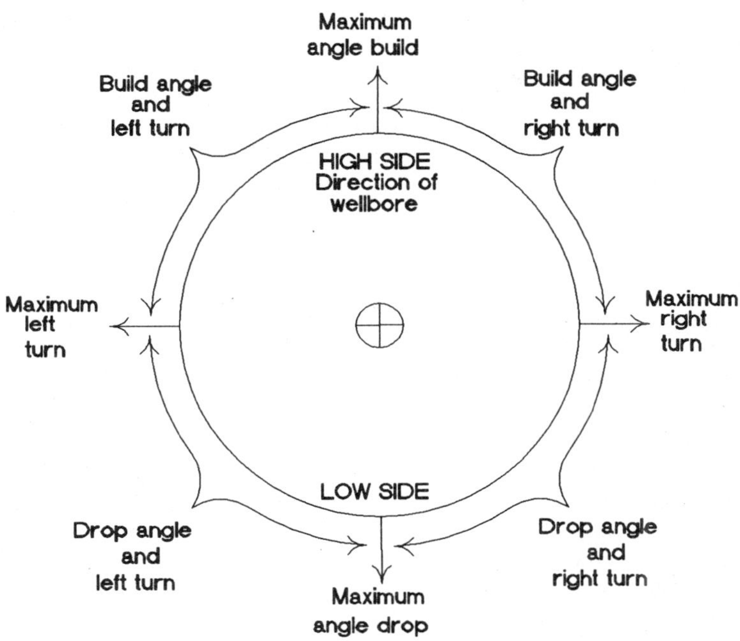

bit to drill to the left of the orientated direction. Reactive torque must be corrected for by turning the assembly in the right direction (clockwise looking downward) during orientation. Corrections range from a few degrees to more than 30°. The amount depends upon various factors, such as the size and length of the drilling assembly, bit weight, rotational speed, and angle of the hole. Reactive torque can be a problem with magnetic single-shot orientation and has been eliminated in later measurement systems. Newer systems measure the direction of the tool face while drilling and provide for immediate corrections.

Empirical tables have values of reactive torque for various conditions. These are used only if no other information is available. Reactive torque should be compensated for during orienting, adding it to other corrections. The tool face is pointed the required number of degrees to the right or clockwise direction (looking downward) of the course of the hole. Then when drilling starts, with weight applied to the bit, reactive torque rotates the assembly to the left or counterclockwise, pointing the tool face in the correct direction. The drift and direction must be measured periodically, ensuring that drilling continues in the correct direction. Changes are made as necessary.

Reactive torque can be calculated for a section of deviated hole after drilling it. Drift and direction are measured from two points some distance apart. The data is entered into a vector analysis computer program. Reactive torque for the section is determined as the approximate difference between projected direction before drilling and the actual results after drilling. This is then applied to the next tool setting, modifying it as necessary. Experienced operators can predict and calculate the correction with good accuracy.

BIT WALK

Bit walk is the change in hole direction due to the rotating bit during drilling. It is caused by the right, clockwise rotation of the bit and by the bit side-cutting action. Bit walk, sometimes called lateral drift, normally causes the hole to turn right (in the clockwise direction looking downward). Severity of the turning action depends upon the type of bit and assembly, bit weight, rotational speed, and formation characteristics.

Bit walk is least in massive, soft formations and increases with increasing formation hardness. Layered formations, especially alternating hard and soft layers, increase bit walk. The build angle increases in the updip direction and decreases downdip. It increases at high angles of inclination and decreases at lower angles. Bottomhole assemblies may affect bit walk; it increases with climbing and dropping assemblies and decreases with packed-hole assemblies. Correct placement of stabilizers reduces bit walk but also may increase the difficulty of controlling hole direction.

Bit walk is not an important factor when using tools that measure drift and direction while drilling. The bit may tend to walk, but it is immediately recognizable, allowing corrective action to be taken before it becomes a problem. Strong, active bit walk can be a problem in both directional and horizontal drilling, sometimes despite the measurement system. Usually, changing to a more aggressive directional assembly corrects the problem.

Bit walk may be compensated for with a lead angle when drilling directionally using the magnetic single-shot for measurements. Lead angle is the number of degrees the drilling assembly must be turned to the left (counterclockwise looking downward) of a direct line to the target during orientation. The hole direction turns to the right during drilling. The lead angle may be calculated or approximated, but normally only after drilling directionally for some distance. Each assembly and bit combination tends to have the same bit walk in the same hole. This provides a correction or guide for subsequent tool runs.

Correcting for total bit walk when first deviating or sidetracking is somewhat common for drilling with rotary assemblies into single targets. A hole curved in the right-hand direction (viewed from the vertical) is drilled into the target. This may not be acceptable for multiple targets. The hole enters the target at a different direction in the horizontal plane than if it had been drilled directionally straight toward the target. This must be resolved when designing the well pattern. Bit walk can be a problem after deviating and while drilling lower hole sections with rotary assemblies. Experienced personnel normally can calculate and predict or estimate it accurately.

DEVIATING ON BOTTOM

Deviating is the procedure for changing the direction of the hole, conventionally at the bottom of the hole. Deviating is done so that the new hole has a different drift and direction from the old upper hole. The term deviation conventionally refers to deviating at the bottom of the hole. Sidetracking often is similar, except that it starts some distance from the bottom of the hole so a lower part of the original hole is sidetracked. The two terms are sometimes used interchangeably. Kicking off is the start of either deviating or sidetracking operations.

Almost any open or cased hole may be deviated on bottom, including both directional and horizontal holes. The diameter of cased holes must be large enough to use standard or slim-hole deviation tools safely. Smaller-sized tools are available but are not as strong, durable, or reliable as larger-sized tools. The deviated hole can be either a directional or horizontal pattern. Holes may be deviated on bottom as a continuation of the planned directional or horizontal drilling program. Special deviation or sidetracking bits are available (see Fig. 3–2).

Either of the three measurement systems may be used depending upon the complexity of the directional or horizontal pattern and operator preference. Steering tool and measurement while drilling (MWD) systems are used in more complex patterns, and MWD is used most often in horizontal holes. The magnetic single-shot

Figure 3–2
Special drill bits
(courtesy of Eastman Christensen, a Baker-Hughes company)

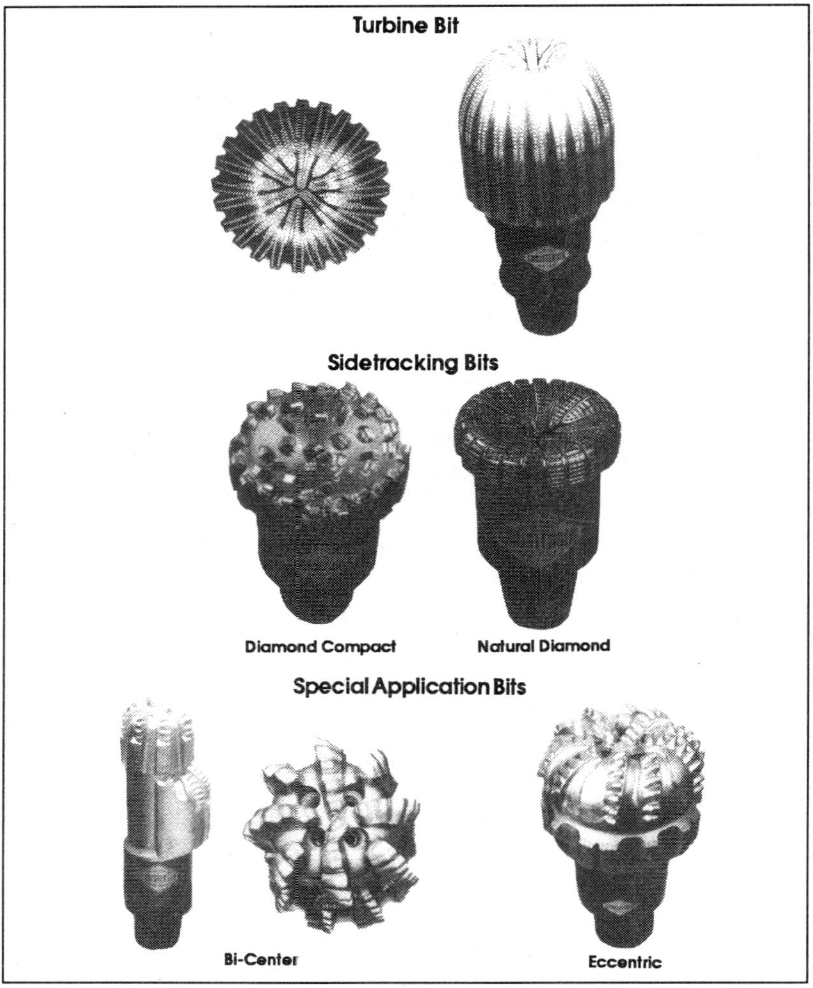

measurement system is explained here for illustrating the procedure for orientation while deviating in the open hole.

OPEN HOLE

A vertical hole is drilled to the kickoff point. (Direction and drift angle are measured while drilling in order to locate the kickoff point.) Some wells may have only drift or angle of inclination measurements. If the cone of uncertainty is acceptable for target limits, deviation begins as planned. Otherwise, the hole is surveyed with a wellbore survey (see Fig. 3–3).

The hole is circulated a full circulation or more to remove all drill cuttings and caving material. In a full circulation, a volume of mud is pumped equivalent to the volume of mud in the hole, without drilling. The hole may be swept with high-gel mud in a viscous sweep for better hole cleaning, if necessary. Normally at least 25 bbl (about 3–5 bbl of mud per inch of hole diameter) are used. Then the drilling assembly is pulled out of the hole. A common deviation motor assembly is built, including a magnetic single-shot orienting sub. The tool face correction (the angular difference between tool face and the indicating magnets) is measured and recorded. The assembly is run into the hole. The kelly is connected and circulated "bottoms up" to remove any formation debris that may have fallen into the hole during tripping. The drillstring is reciprocated periodically with slow rotation during most circulating periods to

Figure 3–3
Deviating on bottom in an open hole

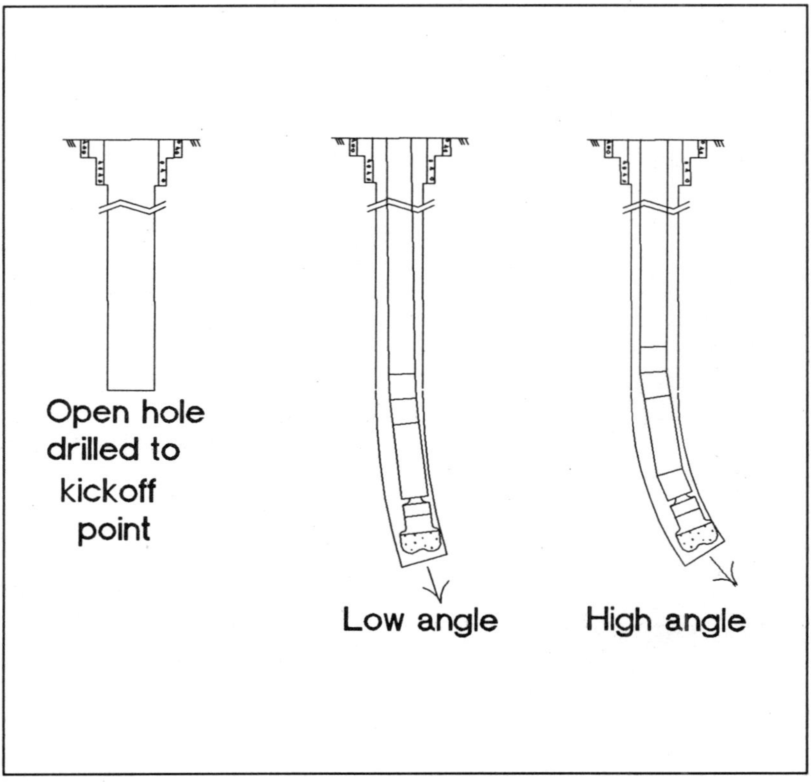

prevent sticking. The clean hole also helps to prevent sticking during the orientation process. The drillstring is stopped with the bit near the bottom of the hole. The kelly is removed and set aside to begin the orientation procedure with the magnetic single-shot.

First the drift and direction of the hole and the assembly tool face are measured. The bit drills in the direction of the tool face (the direction of curvature of the bent sub in the bottomhole assembly [BHA] and opposite the apex of the bend). A magnetic single-shot instrument is lowered inside the drillpipe with a single-strand wireline. The drillstring is left stationary, allowing time for the measuring instruments to come to a complete stop before recording drift, direction, and tool face measurements. The motion sensor generally is better than timer-type instruments here. The measuring instruments are pulled out of the hole and the measurements are observed. It is necessary to ensure that the tool face indicating needle is opposite the indicating magnets in the orienting sub. Additional surveys should be run if needed.

The tool face direction should be corrected for the difference between the tool face and the indicating magnets. Then the measured tool face direction is corrected to true north and this heading or direction is compared to the design direction of the hole. The amount of difference and its horizontal direction determine how many degrees to turn the drillstring and in what direction to point the tool face to the correct kickoff direction.

The drillstring is turned the required amount, allowing for reactive torque and bit walk. The amount of turn at the bottomhole assembly often is less than the turn at the surface because of drag and friction between the drillstring and the walls of the wellbore. The difference is greater in deeper holes, especially deviated, inclined, and crooked holes. This should be corrected for by working the torque down. The drillstring must be prevented from rotating at the surface and reciprocated slowly, moving it up and down several times. This removes the torque in the drillstring so that the amount of turn on bottom is equivalent to the amount of turn at the surface. The bit should be pointed in the correct direction at this time. Another measurement is taken in the previously described manner to verify that the tool face points in the correct direction. If it does not, the drillstring is turned as required, working the torque down and measuring again for confirmation.

The kelly is reconnected and circulation begins, locking the rotary to prevent turning the drillstring. The drillstring is lowered, not allowing it to turn, and a small amount of weight is applied on the formation. The bit, rotated by the motor, begins drilling the deviated hole in the direction of the bend or curve of the bent sub. The weight on the bit is increased until it is in the range recommended for the bit and motor combination. The angle builds at a rate determined by the degrees of bend in the bent sub. Other factors include bit weight, rotational speed, and the formation's tendency to affect the direction of drilling.

About 30 ft or more of deviated hole are drilled, and then drift and direction are measured to verify that the direction of the hole follows the plan. The pump is stopped, and the kelly is disconnected and set back. A joint of drillpipe is connected to the drillstring and lowered so that the deviation assembly is near bottom in the new deviated hole. Drift and direction of the new hole and the tool face direction are recorded with the magnetic single-shot in the manner described. There should be a small increase of angle in the direction of the target. The drillstring must be oriented again if the direction needs to be adjusted. The kelly is connected, the pump started, and deviation drilling resumes. It may be necessary to drill a longer section, up to 50 ft, before the changes of drift and direction are significant. This depends upon the distance between the measuring point and the bottom of the hole and the rate of angle buildup.

Formations affect deviation as noted in Chapter 1. The circulation rate should be reduced, if necessary, in very soft formations. Otherwise the high fluid volume may erode the hole, making angle buildup and directional control less efficient. Hard formations cause reduced penetration rates. Special attention must be given to the bit selection and drilling parameters. Turbines and positive displacement motors have limiting bit weight capacities and may stall under a high load.

Once in a while the angle-build rate may be too low. The first step is to try to increase it by adjusting the bit weight and rotational speed. If this is unsuccessful, the drillstring is pulled out of the hole and the bottomhole directional assembly is modified so that it builds angle at a higher rate. The bent sub is then replaced with another that has a higher degree of bend. Alternately, the bent sub and motor may be replaced with a motor with a bent housing. A bent sub can be added to this for a very aggressive angle-building combination. This will have a very high build rate, such as building curvature for a shorter turn radius horizontal hole. The modified assembly is run into the hole, oriented, and deviation drilling resumes.

At other times, the angle-build rate may be too high. The first step is to try to decrease it by adjusting the bit weight and rotational speed. Then the assembly may be pulled out of the hole if the angle-build rate continues to be too high. It can then be replaced with another that has a smaller angle of bend. This assembly is run back in the hole and deviating resumes. If the angle-build rate is only slightly high, it can be reduced by drilling side-to-side. The drilling assembly is turned a few degrees to one side and drilled for a short time. Then it is turned the same number of degrees toward the opposite side and drilled for a similar period of time. The changes in the sideways directions are small, countering each other, so the net result is a relatively smooth hole with a reduced angle of build. This procedure is not commonly used.

Deviation drilling continues, with periodic measurements and adjustments made as needed until the hole deviates in the correct direction with an established upward curvature. Then the hole is drilled directionally or horizontally by procedures described in Chapter 4 or Chapter 5.

CASED HOLE

A cased hole is deviated on bottom similarly to deviating an open hole. The position of the kickoff point or bottom of the casing is found from prior surveys or a new survey of the hole. This is handled similarly to the open hole situation previously described, except that it is resurveyed with a gyroscopic tool (see Fig. 3–4).

The casing float collar and shoe, if used, are drilled. An open hole section is drilled vertically at least 50 ft and preferably 150 ft or

Figure 3–4
Deviating on bottom in a cased hole

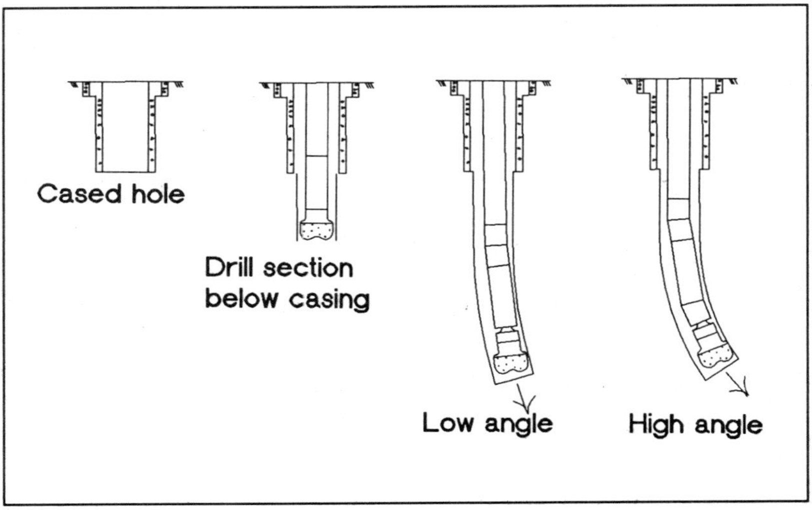

more below the casing. This helps to ensure that the bottom of the casing will not interfere with the deviation operation. The hole is circulated to remove formation cuttings and caving material, and the drilling assembly is pulled out of the hole.

The most common method of deviating in this case is one of direct orientation procedures. The indirect method of orientation is seldom used, as noted, but is applicable in a few situations. Therefore it is described here, referenced to the high side of the hole. Measurements are recorded with the modified magnetic single-shot as previously described.

A drift indicator is run into the open hole on a wireline and the drift and direction of the wellbore are measured. This also gives the high side, which is the same direction as the wellbore. The direction is then corrected to true north. The hole must have about 3 degrees or more of drift, regardless of direction, for measuring the high side accurately when using the indirect method. Most holes commonly have a drift in this range. If not, it may be necessary to drill a short section of deviated hole and measure the drift and direction in the open hole again.

A deviation assembly should be built without nonmagnetic collars or an orienting sub. The assembly is run to a position near the bottom of the hole. A modified magnetic single-shot is lowered on a wireline to the bottom of the deviating assembly. Drift (this also gives the high side) and the direction of the tool face relative to the drift are measured. It must be kept in mind that actual compass directions are not recorded, only angles relative to the high side. The difference between the high side of the hole and the direction of the tool face in degrees is recorded. This difference is added to or subtracted from the direction of the high side of the hole measured with the drift indicator, giving the present compass direction of the tool face. The angular difference between the correct course direction and the present direction of the tool face is calculated. By turning the drillstring the number of degrees equal to this difference, the tool face points in the correct direction and is oriented. The tool face setting is verified with another survey and directional drilling begins.

An example will help clarify the procedure (see Fig. 3–5). First assume that the desired course is north, 30° west. The initial measurement in the open hole has a drift angle of south, 40° east. This is also the direction of the high side of the hole. The measurement in the deviation assembly gave an angular difference of 25° between the high side of the hole and the direction of the tool face. Also, it is west of the high side. Adding 25° to the high side direction of south, 40° east gives a current tool face direction of south, 15°

Figure 3–5
Indirect orientation

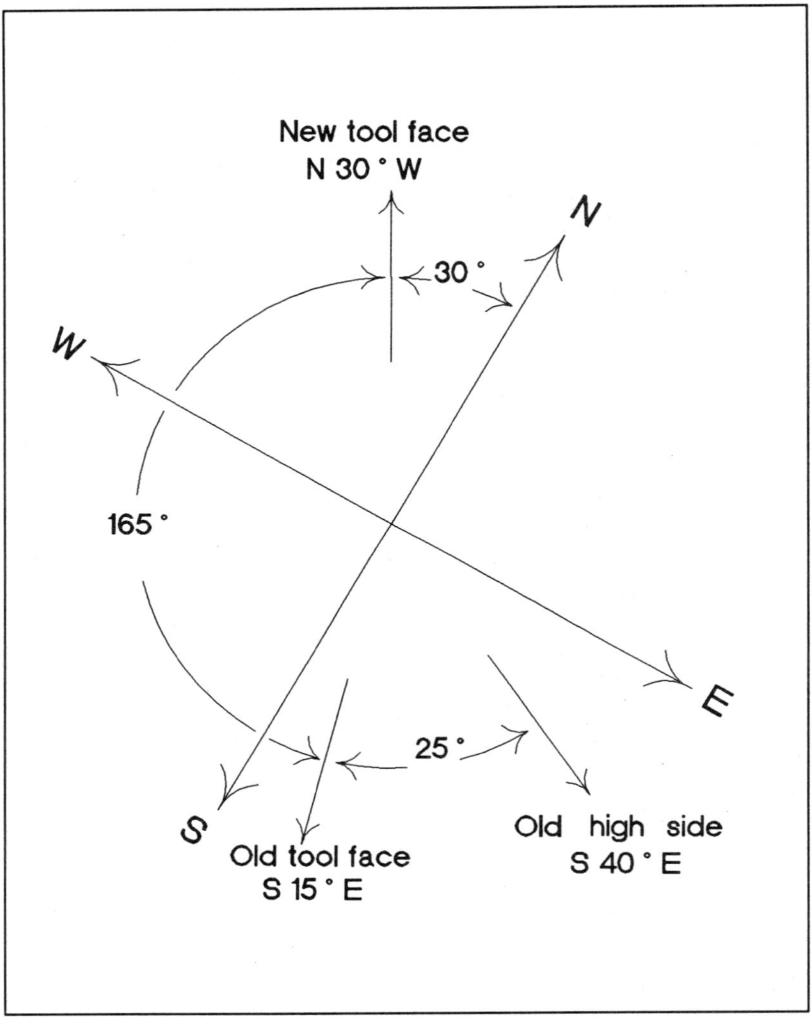

east. This is 165° from the correct course. The tool face is oriented by turning the drillstring 165° clockwise, looking downward. This points the tool face toward the correct direction of north, 30° west.

SIDETRACK PLUG

A sidetrack plug can be placed in open and most cased holes before sidetracking (see Fig. 3–6). A good plug requires correct design and placement, and drilling off a clean top to prevent a failure. The general sidetrack plugging procedure is straightforward, deceptively so, since plugging back frequently is a major sidetracking problem.

The plug serves several purposes. It is the base or seat for deviating tools necessary for sidetracking the original hole. It seals off the lower original hole section, isolating any lost circulation, high pressure, or other troublesome formations exposed in the original wellbore. Otherwise, these formations may adversely affect sidetracking and deviation drilling operations. The plug helps prevent directional tools from entering the original hole while drilling in the sidetracked hole. If this occurs, it is almost impossible to reenter the sidetrack hole, requiring plugging back and sidetracking the original hole again. Additional plugs may be needed in the lower part of the original hole section, subject to good drilling practices and the rules of regulatory agencies having jurisdiction.

Formation hardness, abrasiveness, and stratification may affect sidetracking. It is helpful to sidetrack in medium drillability, massive formations when possible. Normally the precise sidetracking point is not critical, so there is some latitude in selecting it. Prior drilling provides information about formation characteristics. Also, a review of electric logs, penetration rate curves, and similar data helps to find the correct sidetracking point.

Figure 3–6
Sidetrack plug

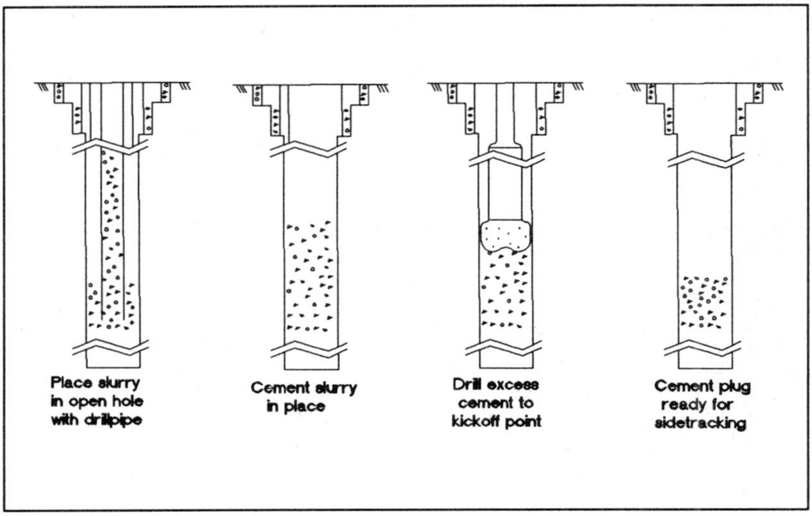

DESIGN

The plug design includes determining the necessary plug length, selecting and designing the type and volume of cement slurry and spacers, and choosing a placement procedure. Plug length is the length of the dressed-off plug that is ready for sidetracking. It is very important to most successful sidetracking operations. The dressed-off plug should be long enough so that the original hole does not interfere with the sidetracked hole. The original and sidetrack holes theoretically separate when the centerlines of the two holes are one hole diameter apart, assuming both have the same diameter. At this separation point, the bit finishes drilling on the plug and begins drilling completely in new formation.

Normal deviation is at a constant angle of buildup of about 2°–2.5°/100 ft. The distance below the kickoff point is less than 50 ft to the separation point for common hole sizes about 6 1/4 in. to 9 7/8 in. This would be a very short plug by field standards. Open holes have been sidetracked above shorter plugs, but they are the exception.

Field experience has clearly established that considerably longer plugs ensure successfully deviating the hole on the first attempt and eliminate the need to set another plug for the reasons described earlier. The recommended dressed-off plug length is at least 200 ft for normal conditions. This requires a slurry plug to be 250–350 ft in length, and 500 ft is not excessive. If there is any doubt, a longer plug should be set.

A shorter plug length should not be selected in order to save the amount of cement needed, to save the extra time required to drill the cement, or to conserve drilled hole. The plug length is found by the horizontal separation required between the deviated hole and the original hole at the bottom of the plug. The plug length is adjusted so that the original and new deviated holes are 3 to 10 bit diameters apart at the bottom of the plug.

A wider separation (longer plugs) should be used in soft, laminated, or naturally fractured formations, and wherever high-pressure formations (saltwater flows, etc.) are exposed in the original hole. THIS IS VERY IMPORTANT. Longer plugs reduce the risk of drilling down the side of a plug or reentering the old hole. Conditions where there is a high risk of this occurring include blind sidetracks, if slurry contamination may occur, and whenever the original hole has been open for a long period of time. Higher angle-build rates should be combined with longer plugs to ensure sidetracking successfully wherever it might be a problem.

The plug slurry should be designed for a high, early maximum compressive strength of 3,000 to 3,500 psi in 24 hrs, using standard design procedures. Class H cement is most commonly used, despite depth, although class A can be used for plugging at shallower depths. Twenty percent to 35% (by volume) of good quality sand should always be added except in very extenuating circumstances. Larger mesh sizes (8–12 or 10–20) should be added if difficult plug problems are anticipated. For most other plugs, 20–40 or 40–60 are used. Finer sizes of 100 mesh or fine "flour" are less preferable but sometimes used. Sand settling in the slurry normally is not a problem. The slurry should be weighted to 15 PPG or 1 PPG more than the mud weight, whichever is heavier. Slurry and mud intermingling due to gravity separation is negligible. Cement slurries with a small swelling tendency may be favorable.

Time spent waiting for the slurry to harden may be minimized by adding accelerators. If conditions require retardation, only a very small amount should be added. A minimum pumping time should be planned for by adding estimated actual mixing and displacement time plus 1 hour. It is important not to design for excessive pumping time. Some types of mud or additives act as retarders and may cause a soft plug. Intermingling and contamination between the mud and slurry may be prevented by separating them with spacers or chemical flushes. Spearhead or lead spacers can be used to clean the walls of the borehole for improved cement-to-formation bonding. The tail in spacers is placed behind the plug. Weight is added to some spacers for deeper plugs set in high-weight mud systems. Spacer volumes normally are somewhat small (5–25 bbls).

It is wise to plan for a cement volume of sufficient size for accurate measurement. Theoretically, a plug of any size can be mixed, pumped, and displaced. But, as a practical matter, there is a minimum usable volume in average-sized holes using standard tools and mixing procedures. It is advisable to always use at least 50 sacks of cement except in extenuating circumstances. The average minimum is about 100 sacks, or 20–30 bbls of slurry depending upon yield. Lesser volumes increase the risk of contaminating the slurry with mud during pumping and displacement. The use of good mixing water is a standard precaution.

Thickening time and compressive strength are tested with the same water to be used for mixing the plug. Initially, slurries are tested for the proper blend of additives with samples of cement taken from the same storage silo containing cement for use on the job. Final tests are run to verify thickening time and compressive strength using cement from the transport truck containing the blended cement and additives. The cement slurry must remain fluid and pumpable during mixing and displacement. After allowing for this, the main criterion for selecting the type of cement and additives is that the plug must have a high, early compressive strength.

PLACEMENT

Placement is the procedure of mixing the slurry and placing it in position in the wellbore. The drillpipe is positioned with the bottom at the same depth as the bottom of the plug and the wellbore is circulated clean. The dry cement is mixed into a slurry with water and additives, normally batch mixed. The spacers are mixed separately. The lead spacer is pumped first, followed by the plug slurry, tail spacer, and displacement fluid (usually mud). Several dry cement samples and wet slurry samples are caught as aids to determine cement hardness and for later analysis if the plug fails. The pressure gauge and densimeter on the cement truck discharge line are monitored. Cement density should be verified by weighing with a mud scale. The plug slurry is displaced to the correct position in the wellbore by balanced or unbalanced columns or by bullheading.

In the balanced columns procedure, the spacers and plug slurry are pumped into the drillpipe as noted. Then a calculated volume of displacement fluid is pumped until the fluid columns inside and outside the drillstring balance. It is necessary to adjust for the density and volume of spacers and slurry and the difference in the density of the displacing fluid and mud in the hole. The drillpipe is pulled slowly out of the cement and normally out of the hole. A wiper plug and catcher separates the slurry or tail spacer and displacing fluid, if used. It gives a positive indication of complete displacement. The balanced column procedure requires careful measurement of fluids, and there is a risk of pulling wet drillpipe.

The underbalanced columns method is similar to balanced columns except that the slurry is deliberately underdisplaced a small amount. Fluid inside the pipe falls a short distance, and the two columns equalize almost immediately. The underbalanced columns method is the easiest procedure to do, and the results generally are favorable. There is minimal risk of pulling wet pipe.

Bullheading is a procedure for pumping the cement slurry directly down the open casing, without drillpipe in the hole. Displacement is accomplished with a volume of fluid calculated to position the top of the plug at the desired point in the hole. The slurry and displacing fluid are separated with a wiper plug if it is not displaced out of the casing. This procedure is seldom used because of the questionable positioning of the plug.

The drillstring is pulled out of the slurry immediately after displacement, excluding bullheading, to prevent sticking. At least 5 to 10 additional stands (3 joints/stand) must be pulled. Pit levels must be monitored while circulating and waiting for the slurry to thicken to immobility. It is necessary to wait for a period of time equivalent to about 2 or 3 thickening times. It is useful to hold low pressure under closed preventers if the system is near balance and there are high-pressure formations open. It is possible to monitor without circulation or pressure if there is a risk of fluid loss in open lost circulation zones. The drillpipe should be moved periodically. Reversing out excess cement normally is not recommended because of the risk of sticking or moving the plug slurry. The remaining drillpipe is pulled out of the hole after the slurry has reached an initial set, usually after waiting the equivalent of 2 or 3 thickening times or longer.

DRESSING OFF THE PLUG

Dressing off the plug is the procedure for drilling the excess cement off the top part of the plug and down to the sidetrack point. A limber rotary assembly is run with a long-tooth soft-formation roller bit, a polycrystalline diamond compact (PDC) bit, or a cement mill. First most of the excess cement is cleaned out while it is soft to save extra time drilling hard cement. One should plan to have cement cleaned out to about 150 ft above the estimated kickoff point before the plug reaches any appreciable compressive strength. THE DRILLING ASSEMBLY SHOULD NEVER BE RUN INTO SOFT (GREEN) CEMENT. This common error causes a difficult sticking situation. It is important to know all the drillstring measurements and the depth to the calculated top of the cement. Channeling, overdisplacement, excess cement, and mixing a lighter weight slurry can cause the cement top to be higher than originally projected. Observe the weight indicator carefully, but do not rely upon it completely, since the pipe may stick before the indicator shows weight. THIS IS VERY IMPORTANT.

Cement-contaminated mud may be a problem requiring one of several actions. The mud may be treated (or pretreated) with chemicals or diluted with water while drilling. The hole may be displaced with old mud or water, which is discarded during or after drilling cement. The hole may be displaced with an inert mud, such as oil mud, that resists contamination by cement. The problem must be handled by standard procedures that depend primarily on the type of mud in the hole and other conditions applicable to the specific well.

The process starts by picking each stand of drillpipe up about 30 ft, ensuring that the drillpipe remains free, when the bit is about 500 ft above the calculated plug top, and is repeated with the following stands. Circulating and reaming down starts at least 250 ft above the calculated top and stops 100–150 ft above the kickoff point, depending upon cement hardness. It is necessary to circulate first in order to condition the mud and then circulate more slowly while waiting on cement (WOC) if the plug has not had time to harden to the correct compressive strength.

The remaining plug is dressed-off in stages using Table 3–1 as a guide to cement hardness. A short section of cement is drilled after the plug slurry has had time to harden and gain sufficient compressive strength. If the cement is hard, Table 3–1 is referred to and then drilling continues to the kickoff point. If the cement is somewhat soft, the drillstring can be picked up a short distance. The hole should be circulated clean and the circulation should continue slowly while waiting for the cement to continue hardening. Waiting time depends upon the relative hardness of the last section of cement drilled. Then the cement hardness should be tested by drilling another short section. The procedure is repeated as necessary until the plug is hard, and then drilling continues to the kickoff point.

Plugs often have hard and soft sections, especially in the open hole. Possible causes are isolated, localized, dilution contamination (probably from mud), extra hydration opposite more porous hole sections, or possibly from improper mixing. Drilling should stop in a harder section. Usually the kickoff point does not have to be at a precise depth and tolerances of 50–100 ft are common.

Table 3–1
Drilling Rate vs. Sidetrack Plug Hardness.

10 ft/hr or 6 min/ft, eqv.—3,500 psi, very hard**
20 ft/hr or 3 min/ft, eqv.—3,000 psi, hard**
30 ft/hr or 2 min/ft, eqv.—2,500 psi, firm**
40 ft/hr or 1.5 min/ft, eqv.—1,500 psi, soft***
50 ft/hr or 1.4 min/ft, eqv.—1,000 psi, very soft****
60 ft/hr or 1 min/ft, eqv.—500 psi, not set****

*Drilling rates in ft/hr or min/ft are related (equivalent to) cement hardness as compressive strength, psi. The data assumes drilling with a

medium-soft formation roller bit, using about 1,000 lbs of bit weight per inch of bit diameter, 50–60 rotary rpm and 1000–1500 psi pump pressure. Normally, tripping the drillstring to run a deviation assembly after dressing off the plug allows additional time for the plug to harden.
** Sufficiently hard for normal sidetracking.
*** Sidetracking very questionable.
**** Drill or circulate out cement and reset plug.

If the cement does not harden within a reasonable period, then it is drilled out to about 20 ft below the bottom of the plug setting depth and another plug is set. Reasonable time depends upon the type of cement, the hole temperature, and many other factors that affect cement hardening. As a guideline, cement should harden a total time of about 200–300% of the calculated hardening time for the desired compressive strength. This completes the plug-back procedure, and the next step is sidetracking.

SIDETRACKING

Sidetracking is the procedure for deviating in an original hole at a point above the bottom and drilling a new hole in a different direction. The new hole may be either directional or horizontal. Sidetracking can be done in almost any open or cased hole, providing the diameter of the hole is of sufficient size to pass standard directional tools. Sidetracking of vertical holes is most common, but almost any directional or horizontal hole can be sidetracked also. Common uses are for bypassing a fish or drilling to another objective located away from the original wellbore. Some holes are sidetracked for the same reasons as deviating. Holes are drilled vertically to obtain information about the formation and then sidetracked for horizontal drilling. Cased holes are sidetracked for similar purposes, especially to permit horizontal drilling, which can increase production.

Various problems may occur during sidetracking. The most common is a failure to deviate because the plug is too soft. This can be corrected by setting a longer plug and dressing it off correctly. Drilling around the plug and back into the original hole, especially in soft formations, is a less common problem that may be corrected by setting a longer plug and sidetracking with a higher build angle. Hard formations may cause special sidetracking problems, especially with soft plugs and sometimes even with good, hard plugs. Some formations are actually harder than the cement plug, so the bit will preferentially drill the plug. This can be corrected by setting the hardest plug possible. It is possible to use a longer plug so that there is more distance for sidetracking. Drilling with reduced weight or possibly "time drilling" with an aggressive deviation assembly also is helpful.

Sidetracking in holes containing oil mud reportedly causes problems, but it shouldn't if the plug slurry is designed and positioned correctly using adequate spacers. Other remedies include setting a longer plug with extra slurry and using a higher sand content.

The main reason for failure to sidetrack successfully (with one plug) is drilling before the slurry hardens properly. Other reasons include using slurry volumes that are too small so that the plug is too short, contaminating the slurry during placement, and not deviating the hole aggressively during kickoff. The underlying reason may be a failure to design a good slurry. It is important to be patient. One can always consider using accelerators, but retarders should be omitted if possible, or only the minimum amount should be used. Most failures require plugging back and sidetracking a second time, an additional and usually unnecessary expense.

It is common to locate the horizontal position of the kickoff point based on measurements taken during drilling. The alternatives are to measure with a wellbore survey or accept target limits within a cone of uncertainty as described in Chapter 1. This usually is acceptable for sidetracking around a fish and for large targets with few limiting hard lines. One of the three measuring systems for measurement and orientation during sidetracking should be used.

OPEN HOLE

Sidetracking in the open hole is accomplished by first setting a cement sidetracking plug and drilling the extra cement to the kickoff point as described earlier in this chapter. The concentric and parallel versions of the steering tool measuring system are described here for measurements and orientation.

For the concentric steering tool measuring system, it is necessary first to build a sidetracking motor assembly, similar to a deviating motor assembly, with a steering tool measurement sub. The tool face correction is measured and recorded, which is the angular difference between the tool face and the indicating magnets. The assembly is lowered to the top of the plug by tripping. The instrument measurement package is lowered inside the drillpipe with a shielded electrical conduit (cable) on the drum of a winch on a cable truck. The instrument package is seated in the measurement sub. A swiveling pressure pack-off is installed on top of the drillpipe and connected to the mud hose. The mud pump is started in order to circulate mud and the bit is rotated with a motor. The direction of the tool face is observed on the data display monitor. It is normal to set required corrections in the surface readout equipment so that it reads the corrected tool face direction. This usually includes the difference between the tool face and the indicating magnets and the correction to true north. The drillstring is turned to point the bit in the required direction and locked to prevent it from rotating (usually by locking the swivel on the traveling block). Drilling of the sidetrack hole begins by lowering the drillstring slowly and applying weight to the bit, increasing the weight slowly until the weight is within the specifications of the motor and bit.

It is important to monitor the drift and direction of the hole and the tool face as drilling continues, orienting again as needed. This is accomplished by unlocking the swivel, turning the drillpipe to the correct direction, and locking the swivel to prevent the drillpipe from rotating. Drilling resumes. Precise measurements are recorded periodically by allowing the deviating tool to stop momentarily.

The next step is to add 1–3 joints of drillpipe to the drillstring when the top of the drillpipe is near the rotary. The mud pump is stopped and the pack-off is disconnected. The instrument package is pulled out of the hole with the winch on the cable truck. The instrument package is lowered into a joint of drillpipe in the mouse hole and the pack-off is connected to the top of the joint. The joint of drill pipe is lifted out of the mouse hole, and another joint is placed in the mouse hole and connected it to the bottom of the first joint. Another joint of drillpipe may be connected if there is sufficient mud hose length and space in the mast. These joints are lifted and connected to the top of the drillstring. The instrument package is lowered inside the drillstring with the cable, and seated in the measurement sub. The pack-off is sealed and the mud pump is started. The sidetracking assembly is oriented, the drillstring locked, and sidetrack drilling resumes.

If the drift angle is not correct, it may be adjusted with different bit weights and rotational speeds. If necessary, it is possible to trip and change the bottomhole assembly as described for deviating in the open hole. The instrument package may be replaced if it fails by pulling it out of the hole from inside the drillstring with the cable on the cable truck and lowering another instrument package into the hole. If the cable parts for any reason, it may be recovered by fishing or pulling the drillstring. Drilling continues, sidetracking the original hole until the new deviated hole is in the correct direction with an established upward curvature. The final step is to drill directionally or horizontally by one of the procedures described in Chapter 4 or Chapter 5.

Sidetracking with the parallel measuring tool system is similar except that the lower part of the cable holding the instrument package is inside the drillstring, and the upper part is outside. The cable passes from inside the pipe to the outside through a side-door sub. The sub contains a seal assembly for sealing around the cable and allowing drilling fluid to be pumped through the drillstring. Normally, the sub is positioned so that the cable is outside the drillpipe in a vertical section of cased hole. These limitations may be modified depending upon specific hole conditions.

For the parallel steering tool measuring system, the first step is to lower a sidetracking motor assembly with a steering tool measurement sub into the hole to the location for the installation of the side-door sub. The instrument package is lowered into the drillpipe and seated in the measurement sub. A side-door sub is connected in the drillstring, the cable is passed through the sub, and it is sealed. The sidetrack assembly is lowered by tripping while simultaneously lowering the cable with the cable truck until the assembly is near the bottom of the hole. The kelly is connected, and the mud pump is started.

Orienting and sidetracking are similar to the procedures for sidetracking with measurement instruments run in the parallel system. Standard drillpipe connections are made. The drillstring and sidetracking assembly are pulled out of the hole and the instrument package is replaced if it fails. Then the assembly is lowered, oriented, and sidetrack drilling begins as described. If the conductor line parts either while drilling or tripping, the connected section is pulled out of the hole, sometimes while pulling the drillstring and fishing when necessary. Drilling continues until the original hole is sidetracked with a new deviated hole drilled in the correct direction with an established upward curvature. Then drilling continues directionally or horizontally by procedures described in Chapter 4 and Chapter 5.

Some sidetracking plugs are too soft to sidetrack by the method described but may be sidetracked by time drilling. The procedure also may apply while sidetracking in very hard formations in which the cement hardness is similar to or less than formation hardness. First a deviation assembly is run with the maximum reasonable angle-build section. The top of the dressed-off plug is touched (tagged) and the assembly is picked up until there is a small amount of bit weight on the plug, usually only noticeable on the sensitive needle or pointer of the weight indicator. The actual weight on the cement top should be almost negligible. The sidetracking assembly is oriented and directional drilling begins.

After about 5 to 20 minutes, the drillstring is lowered a few inches while continuing to rotate the bit and circulating. The procedure continues until about 5–10 ft are drilled. It is important not to use noticeable bit weight in the early part of this procedure. The penetration rate is about 2–4 ft/hr depending upon the bit, plug hardness, and the formation.

The next step is to begin increasing the bit weight very slowly. Normally, the drilling response will show if the bit is sidetracking correctly into the formation or following the old hole. If the procedure is successful sidetracking continues. Otherwise, it is necessary to try it again. If the hole is not successfully sidetracked on the second try, then the soft plug must be drilled out completely and another one set.

CASED HOLE

Cased holes are sidetracked by one of three methods, listed here in order of increasing risk: (a) sidetracking through a milled casing section, (b) whipstocking through a milled casing section, and (c) whipstocking through a casing window. Each has advantages and disadvantages. Measurements are recorded with one of the three measurement systems for orientation depending upon the type of sidetracking. The most applicable method is selected based upon depth, casing size, hole condition, the reason for sidetracking, and operator preference.

Sidetracking fundamentals in cased and open holes generally are similar. However, one major difference is the removal of a section of casing by milling or milling a hole through the side of the casing. Other differences are the methods of plugging back, sidetracking procedures, and some of the tools. The cased wellbore is surveyed with a gyroscopic survey to locate the position of the kickoff point if necessary. The cone of uncertainty may be used if it is applicable.

Sidetracking in cased holes is often a higher risk operation than sidetracking in open holes. Smaller diameter casing requires smaller tools that have less strength than larger tools. Operations are more difficult in smaller holes, and they usually take longer because of the involved procedures and the necessity of removing a section of casing or milling a hole through it. The drillstring may rub and wear against the milled hole through the casing and, in the worst case, become stuck. Special tools like whipstocks may cause operating problems and increase sidetracking costs. There is a risk of the whipstock moving or turning during sidetracking operations or in later deviation drilling after sidetracking. Whipstock sidetracking generally is tedious and time-consuming, involving more trips and equipment, all of which increase the risk of failure. Loss of the hole is not uncommon, requiring sidetracking again. The frequency and severity of problems while sidetracking with a whipstock justify the consideration of redrilling the hole unless the deviation pattern is very simple.

It is important not to sidetrack with a whipstock unless there is strong evidence that it is the best approach, the only reasonable alternative, and is economically justified. A section of casing should be milled in preference to milling a hole through the side of the casing when possible. The length of the deviated section should be limited and low angles of build and drop should be used. Whipstock sidetracking is simple in theory and faster sometimes if it is trouble-free, but problems invariably occur, often severe problems. About the only other advantages of whipstocking are requiring the removal of a shorter section of casing and the ability to omit the sidetrack plug in one procedure. These are not major items if done correctly.

SIDETRACKING THROUGH A MILLED CASING SECTION

Sidetracking through a milled casing section is the most common sidetracking procedure and involves the least risk. It is used for both high and low angles of build, for long sections, and in most other cases. It is a common procedure for reentering an old vertical cased hole for drilling horizontally. Preferred casing size is 7 in. or larger since more operating problems occur while sidetracking inside smaller casing sizes. Larger casing sizes may be necessary if the deviated hole section requires more than one string of casing. Any one of the three measurement systems may be used. The use of measurement-while-drilling (MWD) will be described here for purposes of illustration (see Fig. 3–7).

It is common to plug the lower hole before milling the casing, depending upon formation conditions exposed in the lower hole compared to those in the section where the casing will be removed. A drillable cement retainer is common for plugging. The first step is to connect the retainer to the bottom of the drillpipe and lower it into the hole to the location selected for plugging. This frequently is the same depth as the bottom of the sidetrack plug. Then the retainer is set and mud is pumped through it into the formation, ensuring that the casing is open. The third step is to mix about 25 bbls of cement slurry and pump them into the drill pipe. Mud or

Figure 3–7
Sidetracking a cased hole through a milled section

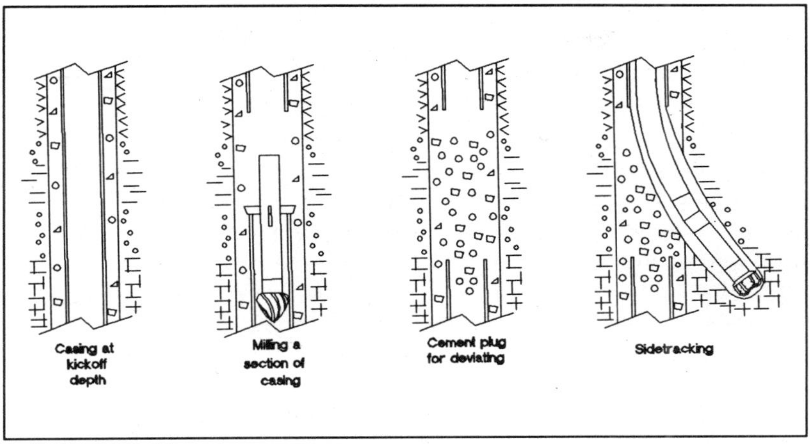

water is pumped behind the slurry and displaced through the retainer into the casing below the retainer. A back pressure valve in the retainer seals and contains pressure below the retainer after pulling the drillpipe. The cement and retainer serve as a double plug.

An alternative procedure is similar except that about half the cement is displaced below the retainer. The next step is to pick up the drillpipe out of the retainer and displace the remaining cement on top of the retainer. This ensures a seal with cement above and below the retainer. Then the drillpipe is pulled out of the hole. Milling casing starts at a point about 20 ft above the projected sidetrack depth. About 60–80 ft of the casing are milled and removed.

A sidetracking cement plug is set as previously described. The bottom of the plug is placed at least 50–100 ft below the bottom of the milled casing section. The plug is extended through the milled section and into the upper casing. After it hardens, the excess cement is drilled or milled so that the top of the plug (kickoff point) is about 20 ft below the top of the milled section of casing. Sidetracking is accomplished in the same general manner as sidetracking in the open hole, allowing for the different type of measuring system, measurement-while-drilling (MWD).

A measurement or instrument sub holds the MWD equipment. The MWD measurement sub is connected in the sidetracking motor assembly. The next step is to measure and record the tool face correction, the angular difference between tool face and the indicating magnets. The assembly is then lowered into the hole. A mud pulse sensor or other type of sensing instrument is installed at the surface, depending upon the MWD system, and the data display monitor also is installed. The kelly is connected to the drillstring and the mud pump is started in order to circulate and to rotate the bit. The direction of the tool face should be checked on the monitor. It is normal to set the corrections in the surface readout equipment for true north and the difference between the tool face and the indicating magnets so that it reads the corrected tool face. Orienting is done by turning the drillstring to point the tool face in the correct direction. Then the rotary is locked to prevent rotating the drillstring. The swivel is locked on the traveling block if the kelly is not used. The drillstring is lowered slowly and sidetrack drilling begins.

Precise measurements are taken periodically for verification by allowing the drillstring to come to a full stop momentarily. The allowances for bit walk and reactive torque may be omitted, since MWD equipment gives the correct direction of the tool face. The direction and orientation are monitored again by turning the drillstring as required. The drillpipe connections are made in the normal manner. The drillstring is lifted out of the hole to replace the MWD equipment if it fails.

It is possible to sidetrack a few cased holes in order to bypass an unrecoverable fish, and the lower part of the hole may be redrilled by blind sidetracking. This is used when it is not necessary to monitor and control the direction of the sidetracked hole. The inclination is still monitored, but sidetracking continues without directional control. Nonmagnetic collars are omitted, and the hole is drilled vertically using regular drift measuring instruments. A hole with junked casing is sidetracked similarly.

Gyroscopic surveys may not be necessary after the new hole is 50–75 ft in a straight-line distance from the nearest section of casing in the original cased hole, depending upon casing size and hole drift. The magnetic influence of the casing is negligible at this distance, so the operator may change to a more economical measuring instrument, depending upon the type of sidetrack hole. Drilling is continued until the new sidetrack hole points in the correct direction and has an established upward curvature. Then directional or horizontal drilling begins using one of the procedures described in Chapter 4 or Chapter 5.

WHIPSTOCKING THROUGH A MILLED CASING SECTION

Whipstocking through a milled casing section is a less common sidetracking procedure in a cased hole. There is less risk as compared to sidetracking by milling a hole (casing window) through the casing wall guided by a whipstock. The lower hole is plugged and about 30–40 ft of casing is removed at the kickoff point by milling. A combination hook-wall packer and whipstock assembly is connected to the bottom of the drillpipe and lowered into the hole. The packer is positioned in the casing a few feet below the bottom of the milled section. The direction of the tool face (the sloping tapered section of the whipstock in this case) is measured, usually with a gyroscopic measuring instrument run on a wireline. The whipstock assembly is turned so that the face points toward the correct direction. Then the packer is set, firmly fixed in place by expanding the packer slips so they grip the inside wall of the casing. The drillpipe is released from the packer and pulled out of the hole.

An alternative procedure has a modified single packer with a whipstock seating device on top. The packer is run and oriented with a gyroscopic tool, making allowances for the tool face correction, depending upon the equipment. The packer is seated and pulled out of the hole. Then the whipstock assembly is run and connected to the seating device on top of the packer. The rotary sidetracking tools are released from the whipstock, usually by shearing a retainer pin.

As the rotary sidetracking assembly is lowered, it guides along the tapered face of the whipstock until it touches the side of the wellbore. A small diameter pilot hole is drilled about 20 ft into the formation, guided by the whipstock, and is drilled in the direction of the whipstock face. The angle of the whipstock, usually 2°–4°, determines the drift angle of the sidetracked hole. The assembly is pulled out of the hole by tripping. A hole opener is connected to the bottom of a limber rotary assembly and lowered into the hole. This tool increases the smaller diameter of the pilot hole section to the regular hole diameter. It does not change the direction or angle of the hole.

Sidetracking is completed with a deviation motor assembly similar to the procedure for sidetracking through a milled section of casing. Gyroscopic surveys are used as needed. Some operators drill out with an angle-building rotary assembly. This relies on the new hole maintaining the direction established by the whipstock until it is beyond the magnetic influence of the casing, and then using magnetic instruments. Sidetrack drilling continues until the hole deviates in the correct direction with an established upward curvature. Then drilling continues directionally or horizontally by a procedure described in Chapter 4 and Chapter 5.

WHIPSTOCKING THROUGH A CASING WINDOW

Whipstocking through a casing window is a less common side-tracking procedure. It is similar to whipstocking through a milled section of casing except that a hole is milled through the casing wall. It is used for drilling short deviated sections with low angles of buildup and inclination. It may be more applicable in smaller sizes of casing. Whipstocking through a casing window has all the disadvantages of whipstocking through a milled casing section and more. There is a higher risk of milling the face of the whipstock or of the mill rolling off the whipstock while milling the window. Tools can stick in the small casing window later while drilling deeper. It is faster than the other methods when successful, but it is a high-risk procedure, generally not recommended (see Fig. 3–8).

A combination hook-wall packer and whipstock starting-mill rotary assembly is connected to the bottom of the drillpipe. It is lowered into the hole to the kickoff point. The whipstock is oriented and the packer is set. The mill assembly is released from the whipstock, the drillstring is lowered and a small diameter hole is milled through the casing wall with a low rotary speed and very little weight on the mill. The assembly is pulled out of the hole and

Figure 3–8
Sidetracking a cased hole through a milled hole

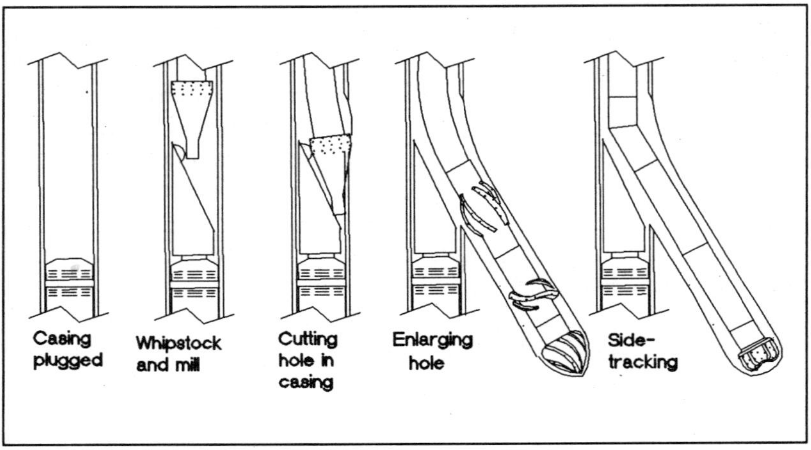

Mills
(courtesy of Eastman Christensen, a Baker-Hughes company)

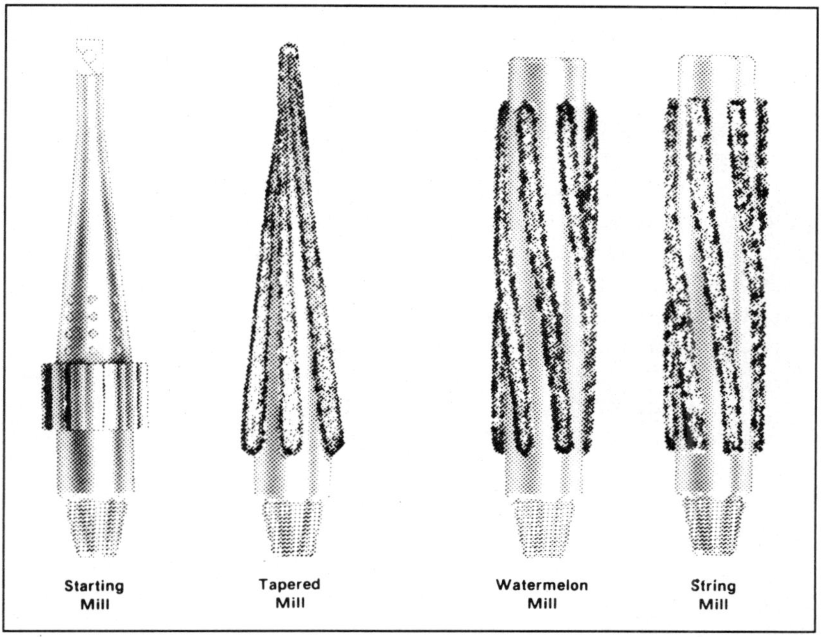

a taper mill is run on bottom with an elliptically shaped reamer mill above it (see Fig. 3–9). The hole is milled in the casing to full gauge, the size of the regular hole, and 10–20 ft are drilled into the formation. This hole is in the direction of the whipstock face at an angle determined by the angle of the whipstock.

The next step is to run a rotary angle-build assembly and drill 30–50 ft, and then pull it out of the hole. A deviation motor assembly is run, and sidetracking is completed similarly to the procedure for whipstocking through a milled casing section. The hole is then drilled directionally. There are various other packer/whipstock combinations and procedures but all are modifications of or are similar to the method described.

MILLING CASING

Milling casing is the procedure of removing a section of casing by milling. The first step is to carefully select the point to start cutting. The lowest joint or part of a joint above the milled section may be loosened or backed off during milling or subsequent sidetracking operations. It is important to ensure that the casing is well cemented in the area of the milled section so that it is firmly fixed in place. This can be verified by reviewing the cement-bond log. It may be necessary to consider perforating and squeezing with cement if the casing is not well cemented. It is necessary to reduce the risk of backing-off by starting milling about 5 feet above a casing collar. This leaves a longer section of casing immediately above the milled section. The extra length improves the chances of a good cement job with less risk of a back-off situation.

The casing is milled with section mills, which have retractable blades (usually three) constructed with a combination of steel and tungsten carbide and designed for milling metal. The section mill is run on a limber bottomhole rotary assembly. The next step is to connect two or three junk subs (boot baskets) in the assembly above the mill to help catch the larger metal cuttings. The milling assembly is lowered into the hole near the top of the section of casing to be milled. The blades or knives are extended by starting the pump and circulating. The assembly is lowered slowly until the extended knives contact a casing collar recess, indicated by a slight decrease in drillstring weight. The assembly is lifted about 3–5 ft and rotated without lowering the assembly so that the knives first cut through the casing wall. The assembly is rotated while being lowered slowly and carefully to start the milling and removal of the casing. At least 50 ft of casing should be milled (preferably 80 ft) depending upon deviation tool requirements.

The assembly is pulled out of the hole if the knives break or become worn. If this is the case, then a new mill, or one with new blades, is lowered and milling resumes until the correct length of casing is removed.

The basic milling procedure is not complex and long sections of casing can be milled. It is possible to mill double sections of casing with a smaller size inside a larger size, and even drill collars have been milled successfully. Milling tool selection is important because a number of tools are available, but some are more efficient, mill faster, and have longer lives than others. Breakage of the section mill knife blades is a common problem, frequently caused by milling too fast, using excessive weight, or not operating the drillstring smoothly. Good mud circulation cools the mill and removes the milled metal cuttings, carrying them to the surface.

Mill cuttings can be very difficult to remove and may require special procedures. They can plug the bit and prevent circulation. Sometimes the cuttings and drilling assembly become magnetized, causing the cuttings to stick to the drill tools. In this case, one or more of the following procedures may help. First a high-viscosity mud, 80–120 API funnel sec, is used, possibly even higher. The pipe is lifted a short distance at 10–30 min. intervals during milling to ensure that cuttings are not stacking up above the mill, because stacked cuttings can cause sticking. All metal cuttings are removed at the flowline and shaker and should be prevented from recirculating in the mud system. Ditch magnets and fine mesh shaker screens (150–180 mesh) should be used. Bit plugging can be a problem, so a mud screen is used at the top of the drillpipe and a back pressure valve float is used above the bit.

The problem of cuttings and sticking to magnetized drill tools may be accelerated when using oil mud. One possible solution is to temporarily change the mud system to a water-base fluid. Double or triple pipe wipers may be used to remove cuttings that stick to the outer walls of the drillpipe. In severe cases, the cuttings can be removed from the inside of the casing with a casing scraper.

OTHER DEVIATION PROCEDURES

Various other deviating systems and procedures are available. These are specialized and only applicable in certain situations.

Wells drilled from platforms are close to each other because of limited space. The wellbores of these directional wells are separated at as shallow depths as possible to prevent drilling into adjacent wellbores. The problem is more critical at shallow depths and less severe at greater depths. Interference between the wellbores is prevented by first separating individual points of entry into the seabed, the bay bottom, or the ocean floor a maximum distance from each other. A common solution is a template, placed on the sea bottom, containing spaces for the number of wells projected to be drilled from the platform. As an example, a 16-well template, 4X4 design, has 4 rows of wells with 4 wells per row, all 10 ft apart.

Vertical conductors extend from the platform to the seabed. A heavy pipe (drive pipe) is driven through each conductor with a pile driver until the bottom of the pipe is deep into the underlying soft sediments and firmly set. Normally, these are driven to the "point of refusal," usually defined as pile driver blows or strokes for moving the drive pipe downward 1 ft. The number of blows per foot at refusal depends upon various factors including the weight of the pile driver hammer, the distance of fall, and the size of drive pipe; it normally is 70–120 blows per foot. It is possible to combine drive pipes and conductors in some situations. Then a hole is drilled vertically through (and below) the drive pipe or conductor with a rotary assembly for several hundred feet so that the drill tools will hang somewhat straight at the platform level. This prevents side forces from interfering with handling the drill tools at the surface. Then the wellbores are separated completely.

Each hole is deviated and drilled radially away from the platform at a low angle-build rate of about 1°/100 ft to a shallow depth of 1,500–2,000 ft. It is important to ensure that the wear bushing in the bottom of the wellhead controls is in good condition, so it must be checked periodically. Finally, the casing is placed in the holes, and the holes are drilled and deviated into their respective targets according to the development plan, usually into areas located radially around the platform.

Jetting or nudging is a procedure for deviating the hole without using conventional directional assemblies. It is most effective in softer formations and for building angles at low build rates. It is a moderately efficient method of directional drilling under favorable conditions but does not have widespread application. The maximum angle buildup is about 0.5°–1.5°/100 ft in holes with low angles of drift. This gives a long, smooth, curved section with seminormal drilling. The procedure is used to gradually separate a group of wells from each other. It is also used for moving the kickoff location in the direction of the target and reducing the angle required in later directional drilling.

A limber rotary assembly is used with a nonmagnetic drill collar and measurement sub above the bit. One large jet and two small jets are placed in a jet-type bit, or one small jet may be plugged. The assembly is lowered to the bottom and oriented so that the large jet nozzle points toward the target. The hole is circulated with drilling fluid at a high rate without rotating the pipe. High jet velocity erodes the formation in the direction of the large jet (usually a few feet of hole, depending on formation hardness). Then the assembly is lifted a short distance and lowered while rotating and reaming the jetted hole. Then 5–10 ft of new hole are drilled in the normal manner. The two procedures are repeated, changing timing and depths drilled and jetted as needed. Drift and direction are measured periodically. The bit is replaced by a jet sub in very soft formations in limited cases (see Fig 3–10).

Pilot hole deviation is another method of deviating in very hard formations, especially in larger diameter holes where normal deviation may be difficult. The procedure is used for building angle at higher rates in a few cases. The procedure is time-consuming and sometimes the results are not as favorable as desired. It is seldom used except in special situations.

The diameter of the lower part of a limber rotary assembly can be reduced and a small-size bit used. A hooligan assembly with a

Figure 3–10
Deviation by jetting or nudging

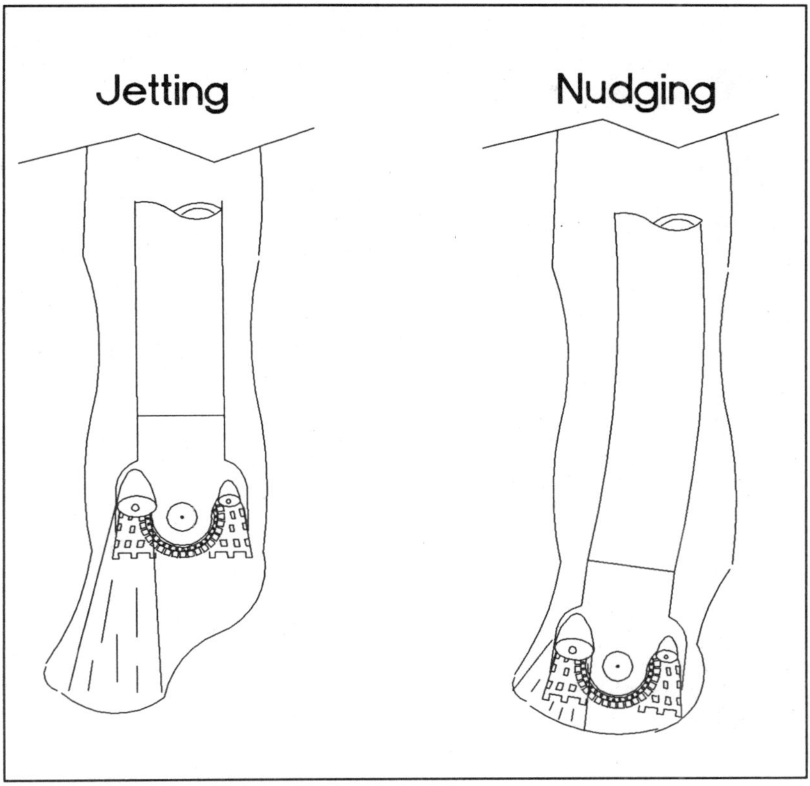

smaller bit can be very effective. The assembly is run to the bottom of the hole and 15–30 ft are drilled with optimal bit weight and rotary speed for building angle. The assembly is pulled and a limber hole opening assembly is run to open the pilot hole to full size. A severe dogleg may be prevented by drilling only a short distance with the small bit and then opening the hole to full gauge. Doglegs cause later keyseating problems. The risk of a dogleg can be reduced by reaming. The procedure should be repeated as necessary. Either rotary or motor assemblies can be used, depending upon specific conditions.

BIBLIOGRAPHY

K. A. Brock and W. S. Cagle. "New Technology Economically Sidetracks Cased Well Bores." *Petroleum Engineer International* (May 1992): 51–54.

W. S. Cagle, et al. "Improved Casing Sidetrack Procedure Now Cuts Wider, Longer Windows." *Petroleum Engineer International* (March 1979): 60–70.

J. M. Dees and W. N. Spradlin, Jr. *Successful Deep Openhole Cement Plugs for the Anadarko Basin.* SPE 10957. Society of Petroleum Engineers/American Institute of Mining and Metallurgical Engineers. New Orleans, LA, September 26–29, 1982.

L. J. Durand, F. A. Samhouri, and D. L. Barthe. "Kicking off in Large-Diameter Holes." *Journal of Petroleum Technology* (October 1982): 2377–2383.

K. K. Millheim. "Proper Application of Directional Drilling Tools Key to Success, Part 2." *Oil & Gas Journal* (November 20, 1978): 155–165.

K. W. Snodgrass. "Fine Well Path for Straight, Curved Conductors." *Oil & Gas Journal* (March 12, 1984): 92–95.

W. Stevenson and W. J. Pike. "Turbodrills Play Major Role in Field Development." *World Oil* (January 1991): 39–41.

G. J. Wilson. "Dog-Leg Control in Directionally Drilled Wells." *Transactions of the American Institute of Mining and Metallurgical Engineers* 240 (1967): 1–107.

CHAPTER 4

DIRECTIONAL DRILLING

SUMMARY

Directional wells are drilled in different patterns at angles to about 60°. Angle is built to about 15° by drilling a smooth curve with the deviation or sidetracking motor assembly. The next step is to complete drilling of the buildup section with the same assembly or with an angle-build rotary assembly. It may be necessary to ream the buildup section to smooth the hole. The straight, inclined section is drilled into the target of a single-bend pattern with a stiff or hold rotary assembly. The end of the straight, inclined section is curved downward with an angle-dropping rotary assembly for a double-bend or *S* pattern. Extended-reach patterns are drilled similarly to the single-bend design but with a longer straight, inclined section.

Complex designs are drilled similarly with additional bends and turns. Slant holes are started at an angle of 30°–45° with a slanted-mast rig and then are drilled directionally, similar to the other holes. Direction, drift, and toolface should be measured periodically while drilling directionally, and correction runs should be made with a deviation motor assembly as required. Operations must be conducted carefully because of increased risks. Major problems include excess drag and torque, casing wear, keyseats, and wall sticking. Fishing procedures can be used to recover tools lost in the holes.

OPERATIONS

Operations in deviated holes include directional drilling of bend-and-run, double-bend, complex, extended-reach, and slant well patterns. Patterns with drift angles of inclination less than about 60° are arbitrarily defined as low angle and are included with directional drilling. This angle is an approximate dividing point because general drilling and completion operations at lower angles are similar to vertical drilling with allowances for deviation. Some of these are reaming, testing, logging, casing, cementing, and completions. Rotary assemblies are efficient at these angles. Operational problems increase substantially at angles greater than about 60°. They are more representative of horizontal drilling and are included in Chapter 5.

These patterns are drilled with standard drilling rigs except that slant holes are drilled with a slant-hole rig (see Chapter 2). Motor assemblies are used for deviation, sidetracking, and correction runs and conventional rotary assemblies are used for all other directional drilling. Motor assemblies can be used, but rotary assemblies are efficient and cost less. Motor assemblies may be slightly more common in offshore drilling as compared to land operations. They are cost-effective offshore because overall operating costs are higher compared to drilling on land and directional equipment costs are a smaller percentage of operating costs. Sometimes motor assemblies may be slightly more efficient in the softer formations that are more common in marine environments.

DRILLING IN DEVIATED HOLES

Drilling in deviated holes is similar to drilling in vertical holes with allowances made for the tools and hole deviation. Bits generally are similar, sometimes with more side-cutting action during deviation. Drilling is slower because most directional assemblies cannot operate efficiently at the higher bit weights and rotary speeds common to vertical drilling. This reduces the penetration rate correspondingly. There are more nondrilling type operations, resulting in an increase in total time. These include measurements, orientation, longer circulation periods for hole cleaning, extra trips for various assemblies, and slower trip time due to extra drag and torque. This nondrilling time should be kept to a minimum. There also are more drilling and related problems, especially in holes with higher inclinations and more complex patterns.

The net result is that less time is spent drilling; directional drilling is slower and has more risk of failures. Careful planning is essential. For example, making a correction with a deviation assembly after drilling with rotary assemblies takes one or two days and can be eliminated by predicting bit walk accurately.

The well plat, developed while planning and designing the well, displays the well plan in horizontal and vertical cross-sectional views at convenient scales. Surveys and measurements are plotted on a copy of the well plat during drilling, using the same scales. This provides a good visual comparison of the planned and final well paths. It is important to always retain all measurements as part of the permanent well record.

Regarding the well plat, the well literally is drilled in three dimensions, and the horizontal and vertical cross sections are in two dimensions. Each cross section individually assumes measurements in a plane. This is not always the case because of small variances in the wellbore. Measurements made while drilling are not always exactly in the same plane but can be projected correctly on the plane. Normally, directional differences less than 8°–10° are insignificant for illustrative purposes but must be provided for in order to drill in the correct direction.

It is important to always maintain a good record of the type of deviating assembly. This includes details on the positioning of diverting tools, stabilizer spacing, reactive torque, bit walk, and overall assembly performance. These records help determine the design of later assemblies. They also aid in finding corrections for bit walk and reactive torque for orientation.

Drilling with a positive displacement motor, sometimes called PDM, is somewhat different than drilling with a regular rotary. One of the main differences is the method of noting bit weight. The weight indicator is a primary tool for observing bit weight in normal rotary drilling. Most indicators do not have sufficient sensitivity to allow adding bit weight while drilling with motors. Motors operate in specific pressure ranges, so the mud pressure gauge is the best guide to drilling. Normally, higher pressure means higher motor torque output. Excessive pressure is equivalent to increased weight, which can stall or possibly damage the motor and may affect the direction of the tool face. Consequently, bit weights must be controlled carefully. High temperatures may affect seals. Motors are subject to wear and failure, and hydraulics are important in order to operate motors correctly.

BIT/FORMATION EFFECTS ON HOLE DIRECTION

Bits and formations influence the direction of the wellbore. The action is site-specific and dependent upon factors in the individual well. A limber assembly, without stabilizers, drills downward through massive formations, such as a sandstone or massive shale, in a slow right-hand spiral (clockwise looking downward). Right-hand rotation and the side-cutting action of the bit on the low side of the hole cause this, and the action is related to bit walk. The spiraling effect ranges from a few horizontal degrees per thousand vertical feet drilled in pure massive formations to more than 20°/1,000 ft in layered, harder formations.

The spiraling effect increases with increasing formation hardness and in layered formations, especially with layers of alternating hardness. It also increases with higher drilling weight and when drilling with bits that have a more aggressive cutting structure. This includes strong side-cutting action and increased roller-bit cone offset. The action also increases with assemblies that have less stabilization. The spiraling effect may be partially or completely obscured where the formations exert a strong angle-building tendency.

Bits tend to drill updip in dipping formations where the relative angle between the axis of the bit and the formation dip is less than about 45°. They drill downdip at higher relative angles. There is less tendency to change the direction of the hole while drilling with stabilized assemblies such as stiff or hold assemblies. Faster bit rotation reduces the tendency to change hole direction. The net result is that these actions cause changes in the direction of the hole in during drilling. Some special 2- and 4-cone bits, designed for directional drilling, reportedly reduce right-hand walk and may cause left-hand bit walk. The bits have a special standardized code number ending in the numeral 8. Understanding the interrelated influence of formations and bit action on the well path helps the operator judge which type of assembly and bit to use and how to operate them.

BIT SELECTION

The factors involved in bit selection are unquestionably important and sometimes intangible items that affect penetration rate. Bit selection depends upon the type of formations, drift, bit weight and rotary speed, hydraulics, and operating conditions. Operating conditions include hole size, rotational speed, weight on bit, torque and type of drilling assembly. Drilling with the correct bit and in the correct manner is critical to all efficient drilling operations. It is especially important in directional and horizontal drilling because of the limited operating conditions. Bit selection as described here also applies to horizontal drilling, except where noted.

It helps to know which bits are more commonly used in the area. A systematic analysis of all factors is the recommended approach; this ranges from an analysis of bit records on offset wells to a detailed study of bit performance in the area. This saves experimentation when it is not clear which is the best bit to use. Bit performance records on nearby reference wells are very good guides. The best bit for straight-hole drilling frequently is the best bit for directional drilling.

Modern drill bits are precision tools and highly reliable when operated within design limits. There are general guides to selecting the correct bit. (Bits are illustrated in Figs. 4–1 and 4–2 for additional clarification of the following guides.)

1. *Give preference to solid body, one-piece, fixed-cutter drag bits such as polycrystalline diamond or regular diamond bits where applicable. These bits do not have moving parts, so the risk of leaving junk in the hole when the bit fails is small. They are durable and withstand high speed rotation (see Fig. 4–1).*
2. *Polycrystalline diamond bits are preferred especially for their characteristic ability to drill quickly and their longevity in applicable formations. They are able to drill long sections of hole. For example, a 12 1/4 in. PDC bit driven by a downhole turbine drilled almost 18,000 ft at about 70 ft/hr from a platform in the North Sea.*
3. *Select bits reinforced in the gauge and shank area, such as those with tungsten carbide inserts. Side loading in directional and horizontal drilling causes extra wear on the side of the bit. Reinforced gauge areas improve performance and reduce the risk of drilling an undergauged hole. These bits are very applicable when drilling in the steerable mode with high bit (not cone) offset.*
4. *Select bits with aggressive side-cutting structure for efficient and rapid changes in direction and angle such as sidetracking and correction runs. This includes moderately aggressive cutting structure and gauge protection.*
5. *Select bits with minimum side-cutting action for drilling straight such as straight, inclined hole sections.*
6. *Roller bits are commonly used. Select a premium grade for a longer bearing life.*
7. *Select roller bits with maximum bit-cone offset for increased penetration rate. This may reduce bit life and cause increased bit walk (see Fig. 4–2).*

Figure 4–1
Solid-body, one-piece bits
(courtesy of Eastman Christensen, a Baker-Hughes company)

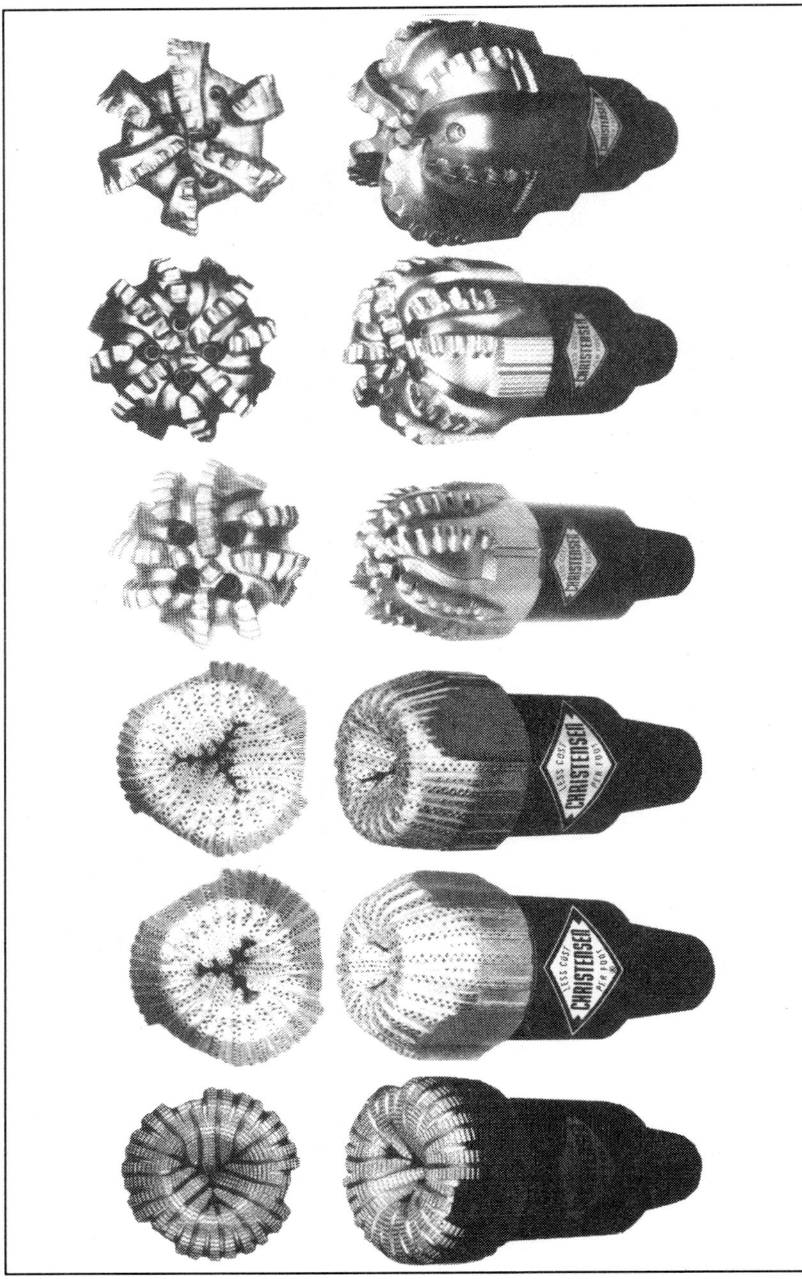

Figure 4–2
Roller bits
(courtesy of Hughes Tool Company, a Baker-Hughes company)

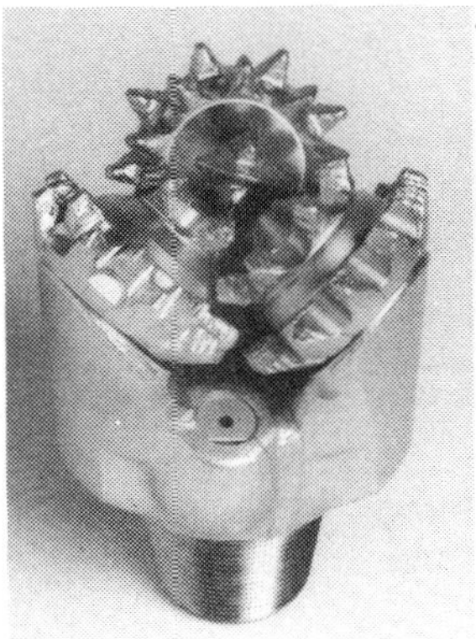

Soft formations

Medium formations

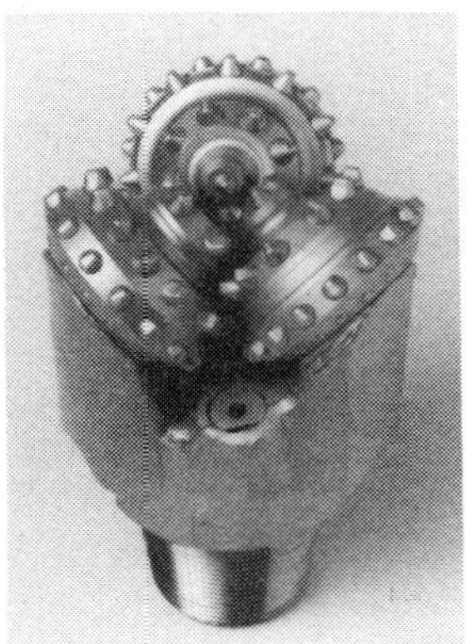

Hard formations

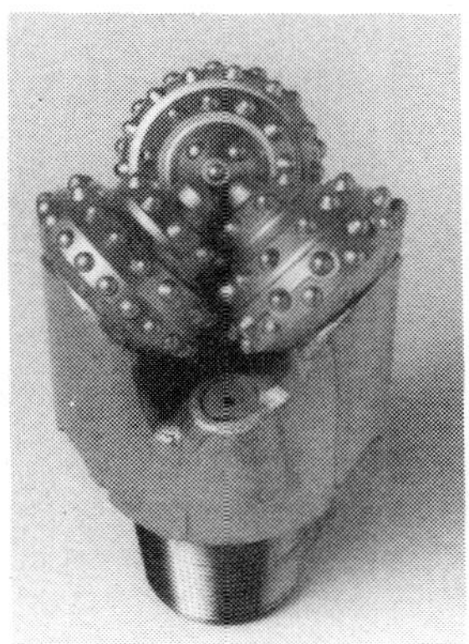

Very hard and abrasive formations

8. *Select diamond bits with larger stones (2–4 stones per carat) in softer formations and smaller stones (3–5 stones per carat) in very hard formations.*
9. *The increased rotation rate of the motors and turbines accelerates bit wear compared to conventional rotary drilling. Allow for this when evaluating bit life. Unnecessary correction runs and inefficient drilling practices increase wear. Bits with higher offset wear faster. Excess circulation may cause roller-bit bearing and seal wear if the bit is off bottom where the cones can spin freely.*
10. *Fixed cutter bits with shorter gauge lengths and aggressive side cutting increase steerability.*
11. *Use near-bit stabilization with solid-body bits and in other cases whenever applicable.*
12. *Observe good bit hydraulics. Bits for directional and horizon tal drilling may have slightly smaller jets because of the pressure drop across the mud motor.*
13. *Drill straight, inclined sections with less aggressive side-cutting action.*
14. *Decrease torque and rough running characteristics with less side-cutting action, smooth gauge protection, and a minimum number (NOT SIZE) of mud courses.*

DRILLING FLUID SELECTION

Drilling fluid (mud) is an important part of any drilling operation. It is of major importance in directional and horizontal drilling in order to maximize the penetration rate and because of the higher risk of hole problems. Higher quality muds often reduce problems caused by drag, torque, keyseating, and wall sticking due to hole curvature and deviation. Drilling mud is a major expense in directional and horizontal drilling, so the mud must be selected carefully.

Selection of the correct mud is similar in all types of drilling but requires special consideration in directional and horizontal drilling. It depends primarily upon formation conditions and the complexity of the directional pattern in the specific well under consideration. Major considerations in mud selection include the type of formation, how the formations affect hole conditions, drilling rates, difficulty in controlling the direction and angle of the hole, and complexity of the directional and horizontal pattern. Other factors relating to the mud are treatment requirements, stability at higher temperatures, resistance to contaminants such as cement contamination, solids transport ability, and ease of separating the solids at the surface.

Almost all types of mud have been used at one time or another in directional and horizontal drilling, even including air to a limited extent. Some muds are better in certain areas and different muds in others, mainly dependent upon the formation conditions. It is common to deviate or sidetrack in the open hole with the same mud used for drilling the vertical section. Then the mud is upgraded while drilling deeper by dispersing an undispersed mud system or improving the quality by changing its physical and chemical characteristics.

A reliable guide is often the kind of mud successfully used for drilling other wells in adjacent areas, preferably directional and horizontal wells but also vertical holes. It is also important to be aware of recent developments. Common mud for the area is used for drilling simpler patterns such as a single, low-angle bend and short, straight section in shallow- and medium-depth holes with minimum formation problems.

More complicated patterns can be drilled with the same type of mud used in areas with fewer formation problems. The mud quality can be upgraded if formation problems occur. Involved programs such as a double-bend or a higher angle single-bend with a longer straight, inclined section in deeper holes may require a better type and sometimes a higher quality mud. Use the best type and highest quality mud for complicated patterns in areas with formation problems.

Oil muds are highly recommended for complex directional programs and drilling in areas with severe hole problems, excluding loss circulation. They are very good for drilling complex directional and horizontal patterns, especially in areas with formation problems. They help in other operations including fishing, difficult sidetracking, and drilling in holes with high drag and torque.

Oil mud has many advantages. It is stable, has good flow properties even at high temperatures, and is easy to maintain. High lubricity reduces drag and torque. Oil mud will not appreciably damage zones containing oil and gas. It is inert and resists contamination from most common formation fluids and when drilling cement. It does not cause hydration, so there is less accumulation of low-gravity solids. Bentonitic shale formations can be stabilized while drilling with oil mud by increasing the salt content in the water phase to a high level. There is also less risk of sticking, especially differential pressure sticking.

However, oil muds do have disadvantages. Gas solubility can create problems in special conditions, and oil muds require special handling equipment, including oil-resistant clothing for personnel.

Oil muds often cost more than other muds, but part of the mud can be salvaged at completion to reduce cost. Often the higher cost is more than justified by the reduction in drilling problems. There may be disposal problems after drilling because of the diesel oil base. Environmental restrictions can be severe, especially in offshore operations. Mineral oil may be used as a base and is more environmentally acceptable than diesel oil, but it is more costly and may not be as useful.

HYDRAULICS

Hydraulics is the general term for expressing pressure drop of the drilling mud throughout the circulating system. Pressure drop relates directly to hydraulic horsepower, so calculating hydraulics shows where pump horsepower is used. Hydraulics are calculated by proprietary computer programs that use data from the wellbore, mud characteristics, and some equipment. Hydraulics are calculated when designing directional and horizontal drilling programs (see Table 4–1). This determines the size of the pumping equipment required. Directional and horizontal drilling may require larger mud pumps compared to vertical drilling because of the extra hydraulic horsepower required for the motor or turbine. This is not normally a problem.

The pressure drop across bit nozzles and turbines or positive displacement motors shows the amount of hydraulic horsepower used in each case. Bit hydraulics help find the correct nozzle sizes for jet bits. Slightly smaller bit nozzles may be required to allow for adequate pressure drop across the positive displacement motor or turbine. Similarly, calculations show where losses occur. Drillpipe that has a small inside diameter often has high losses, so it may be necessary to replace it with a larger size of pipe.

MUD-HANDLING EQUIPMENT

Mud-handling equipment is the surface equipment used to handle, store, mix, and treat the mud as well as to remove solids. The equipment used for most direction and horizontal wells is similar to that used in vertical drilling. Directional and horizontal wells often place greater demands upon this type of equipment, especially for mud treatment and solids removal. Higher quality muds are more common, so adequate-sized mud tanks with good treating and mixing facilities are necessary. Oil mud may require additional equipment, such as drain pans under the floor and good valves where water lines connect to the mud system. A good trip tank is highly recommended; gravity types usually do very well.

Muds used in directional and horizontal drilling often have high mud solids. These are concentrations of drill cuttings, caving material that falls into the hole, sand, and other small particles of the formations. Mud solids range in size from large particles such as drill cuttings to very small, low-gravity solids. Low-gravity solids are drill cuttings and similar material crushed and pulverized by the circulating mud and the drillstring moving in the hole. These smaller particles are more difficult to remove and the difficulty increases with decreasing particle size. There are higher concentrations of these in directional and horizontal drilling for several reasons. The drillstring commonly lies on the low side of the hole during both drilling and tripping, grinding particles into small sizes. It also is difficult to remove these from the mud system, as described later in this chapter. Mud solids degrade the mud, resulting in deteriorating properties such as high gels and viscosity. These in turn can cause operating problems such as reduced cleaning, higher circulating pressures, possible lost circulation, and, in severe cases, sticking of the drillstring. High sand content causes increased wear on the mud pumps and circulating system.

Solids removal equipment removes mud solids. Therefore, good equipment is vital in all efficient drilling operations. It often is very important in directional drilling and can be especially critical in horizontal drilling. Shale shakers are the first step in removing mud solids. These range from single or twin units in single- or double-deck models. High-speed shakers are efficient. Fine mesh screens (at least 80 mesh or smaller) should be used, and even finer mesh on twin shakers. Maximum flow time over the screen must be provided. Desanders remove sand grains from the mud, and finer particles are removed with centrifuges (mud cyclones). Ditch magnets remove iron cuttings. These also indicate pipe and casing wear. Additional or more efficient solids removal equipment should be installed if necessary.

SINGLE-BEND

Single-bend or bend-and-run patterns are common in directional drilling. The drilling program contains the well pattern, including build angles and lengths of the curved and straight sections. The first step is to deviate, drill the buildup section to the required angle, and then drill a straight, inclined section into the target. Reaming and correction runs are made as needed. Steering

Table 4–1
Sample Hydraulics Calculations.
(courtesy Drilex Systems, Inc.)

Surface Data

Surface Equipment Type	:	3
Maximum Standpipe Pressure	:	4000

Hole Data

Section	ID (in.)	Depth (ft)
20" Surf. Casing	19.1240	120.00
13 3/8" Casing	12.7150	6380.00
9 5/8" Casing	9.0630	7205.00
7" Liner	6.5380	8005.00
Open Hole	6.0000	Balance

BHA Data

Section	OD (in.)	ID (in.)	Length (ft)
Drill Pipe	5.5000	4.7780	Balance
HWDP	5.0000	3.0000	93.00
Drill Collar	6.7500	2.2500	150.00
HWDP	5.0000	3.0000	2305.00
Drill Pipe	3.5000	2.7640	1538.00
HWDP	5.0000	3.0000	93.00
NMDC	4.7500	2.5000	31.00
MWD	4.7500		30.00
D475	4.7500		22.38
Fixed Bit	6.0000		1.00

Case Data

Mud Weight (ppg):	9.5	9.5	9.5	9.5
Plastic Viscosity (Cp):	10.0	10.0	10.0	10.0
Yield Point (lbs/100 ft²):	15.0	15.0	15.0	15.0

Pressure Losses (psi)

Depth (ft)		**8004.0**	**9719.0**	**8004.0**	**9719.0**
Flow Rate (gpm)		**200.0**	**200.0**	**250.0**	**250.0**
Surface Equipment:		7.59	7.59	11.49	11.49
Drill Pipe:	Bore:	22.50	32.82	33.78	49.26
	Annulus:	7.14	10.41	10.72	15.63
HWDP:	Bore:	5.27	5.27	7.91	7.91
	Annulus:	0.13	0.13	0.19	0.19
Drill Collar:	Bore:	34.03	34.03	51.08	51.08
	Annulus:	0.90	0.90	1.35	1.35
HWDP:	Bore:	130.68	130.68	196.15	196.15
	Annulus:	3.16	31.02	4.75	46.56
Drill Pipe:	Bore:	129.42	129.42	194.26	194.26
	Annulus:	4.37	18.48	6.56	27.74
HWDP:	Bore:	5.27	5.27	7.91	7.91
	Annulus:	3.37	13.38	5.06	20.08
NMDC:	Bore:	4.23	4.23	6.35	6.35
	Annulus:	0.74	2.38	1.12	3.57
MWD:	Bore:	250.00	250.00	250.00	250.00
	Annulus:	0.72	2.30	1.08	3.46
DRILEX D475 Motor:					
	Off Bottom:	125.00	125.00	150.00	150.00
	At 860.0 ft-lbs torque:	575.00	575.00	600.00	600.00
	Annulus:	0.54	1.72	0.81	2.58
Across Fixed Bit:		287.70	287.70	326.37	326.37
Total Off Bottom:		1023.50	1092.74	1268.03	1371.96
Total On Bottom:		1473.50	1542.74	1718.03	1821.96

Annular and Critical Velocities (fpm)

Depth (ft):		**8004.0**	**9719.0**	**8004.0**	**9719.0**
Flow Rate (gpm):		**200.0**	**200.0**	**250.0**	**250.0**
Drill Pipe in 9 5/8" Casing					
	Annular:	94.47	94.47	118.09	118.09
	Critical:	268.19	268.19	268.19	268.19
	Flow Regime:	Laminar	Laminar	Laminar	Laminar
HWDP in 9 5/8" Casing					
	Annular:	85.79	85.79	107.24	107.24
	Critical:	265.67	265.67	265.67	265.67
	Flow Regime:	Laminar	Laminar	Laminar	Laminar

Drill Collar In 9 5/8" Casing					
	Annular:	134.02	134.02	167.53	167.53
	Critical:	279.55	279.55	279.55	279.55
	Flow Regime:	Laminar	Laminar	Laminar	Laminar
HWDP In 9 5/8" Casing					
	Annular:	85.79	85.79	107.24	107.24
	Critical:	265.67	265.67	265.67	265.67
	Flow Regime:	Laminar	Laminar	Laminar	Laminar
HWDP In 7" Liner					
	Annular:	N/A	276.24	N/A	345.30
	Critical:		296.59		296.59
	Flow Regime:		Laminar		Turbulent
Drill Pipe In 9 5/8" Casing					
	Annular:	70.14	N/A	87.68	N/A
	Critical:	260.88		260.88	
	Flow Regime:	Laminar		Laminar	
Drill Pipe In 7" Liner					
	Annular:	160.75	160.75	200.93	200.93
	Critical:	271.78	271.78	271.78	271.78
	Flow Regime:	Laminar	Laminar	Laminar	Laminar
Drill Pipe In Open Hole					
	Annular:	N/A	206.40	N/A	258.00
	Critical:		277.09		277.09
	Flow Regime:		Laminar		Laminar
HWDP In 7" Liner					
	Annular:	276.24	N/A	345.30	N/A
	Critical:	296.59		296.59	
	Flow Regime:	Laminar		Turbulent	
HWDP In Open Hole					
	Annular:	N/A	445.64	N/A	557.05
	Critical:		325.72		325.72
	Flow Regime:		Turbulent		Turbulent
NMDC In 7" Liner					
	Annular:	242.88	N/A	303.60	N/A
	Critical:	289.38		289.38	
	Flow Regime:	Laminar		Turbulent	
NMDC In Open Hole					
	Annular:	N/A	364.80	N/A	456.00
	Critical:		308.81		308.81
	Flow Regime:		Turbulent		Turbulent
MWD In 7" Liner					
	Annular:	242.88	N/A	303.60	N/A
	Critical:	289.38		289.38	
	Flow Regime:	Laminar		Turbulent	
MWD In Open Hole					
	Annular:	N/A	364.80	N/A	456.00
	Critical:		308.81		308.81
	Flow Regime:		Turbulent		Turbulent
D475 In 7" Liner					
	Annular:	242.88	N/A	303.60	N/A
	Critical:	289.38		289.38	
	Flow Regime:	Laminar		Turbulent	
D475 In Open Hole					
	Annular:	N/A	364.80	N/A	456.00
	Critical:		308.81		308.81
	Flow Regime:		Turbulent		Turbulent

Bit Performance

Depth (ft): **Flow Rate (gpm):**	**8004.0** **200.0**	**9719.0** **200.0**	**8004.0** **250.0**	**9719.0** **250.0**
Bit TFA (in.²)	0.3313 (12-12-12)	0.3313 (12-12-12)	0.3889 (13-13-13)	0.3889 (13-13-13)
Nozzle Velocity (fps):	184.07	184.07	196.05	196.05
Impact Force (lbs):	171.84	171.84	228.79	228.79
Hydraulic Horsepower (hp):	31.89	31.89	45.22	45.22

tools are commonly used with deviation assemblies and magnetic single-shot measuring equipment is used with the rotary assemblies, since they cannot be steered in the horizontal direction. Other measuring equipment can be used also (see Fig 4–3).

BUILDING ANGLE

The angle buildup section is drilled after deviating or sidetracking. It is normal to continue building angle with the same deviation motor assembly used for deviating or sidetracking in the open hole style. Drilling continues in the direction of the target, building angle in a smooth, upward curve. The drift angle is built up to at least 10°–15°, or about one-third of the total angle required, whichever buildup is larger. A complete separation from an old hole must be ensured if sidetracking. Normally this angle is sufficient to establish direction and curvature. A lower angle is used sometimes if sidetracking to bypass a fish or junked hole as described in a later section of this chapter. The remaining buildup section is drilled with an angle-build rotary or motor assembly using either of the three measurement systems.

Angle-building rotary assemblies are often the most efficient, economical assemblies in simple patterns, and they normally drill faster than deviation motor assemblies. There is less risk of sticking and other downhole problems while drilling and a better chance of releasing stuck angle-build rotary assemblies compared to deviation assemblies. Predicting lead angles and correcting for bit walk can be minor obstacles, but most experienced personnel handle these correctly. A motor assembly is used for more complex patterns and building angle at a higher rate. It also is used when

Figure 4–3
Drilling a single-bend directional well

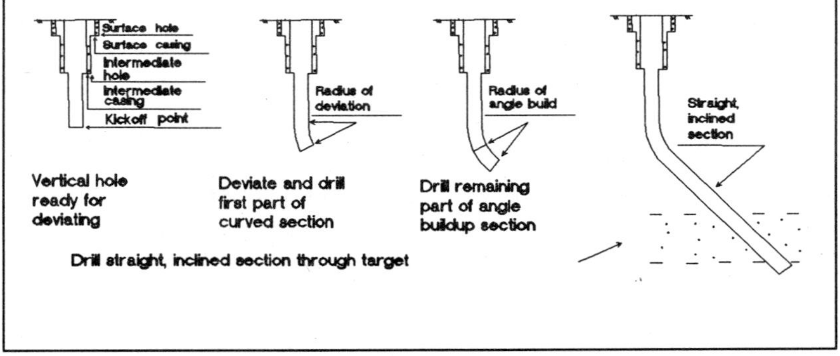

drilling formations that tend to cause the hole to drill in other, incorrect directions.

The deviation or sidetracking assembly is pulled and an angle-build rotary assembly with a single-shot measurement sub is run into the hole. The next step is to find turn and direction, orient with these corrections, and begin drilling, continuing angle buildup. Drift and direction are measured initially at intervals of about 30 ft. Then the distance between measurements is increased to about 60 ft, and later 90 ft, as tool performance becomes more predictable. Minor changes in the angle may be made by changing either the rotary speed or the bit weight.

Trips are made as necessary for changing stabilizer positions on the bottomhole assembly to increase or decrease the rate of angle build. Other reasons for tripping include replacing worn bits or stabilizers or otherwise changing equipment on the bottomhole assembly. The assembly is pulled in the normal manner, the changes are made, and it is rerun into the hole. Measurement surveys are included with trips when possible to save time. After the new assembly is on bottom, the turn and direction are found and these corrections are used for orientation before drilling begins again.

The direction of the hole may change from the planned course, often due to the formations or bit walk. Sometimes the rotary assembly does not build angle at a sufficiently high rate. In these cases, a deviation motor assembly is run and deviation drilling continues with the procedures described later in the section about correction runs. Another rotary assembly is run after completing the correction run or drilling is continued with the deviation motor assembly as an angle-build motor assembly. Drilling with the angle-build motor assembly is similar to deviating and sidetracking. Orienting, measuring drift, direction, and tool face, and tripping continue as necessary. If the assembly does not build angle correctly, it is pulled and another one is run that is either more or less aggressive, depending upon the situation. The hole may be drilled by either procedure, continuing to build angle until the drift angle is about 90% of the well plan.

REAMING

Reaming is a procedure for smoothing out and removing irregularities in the wellbore so that other tools can pass freely. It restores undergauge holes to full gauge, removes keyseats, and reduces or wipes out excessive hole curvature over short intervals near the ends of long curved sections (see Fig. 4–4).

Figure 4–4
Reaming and accidental sidetracking

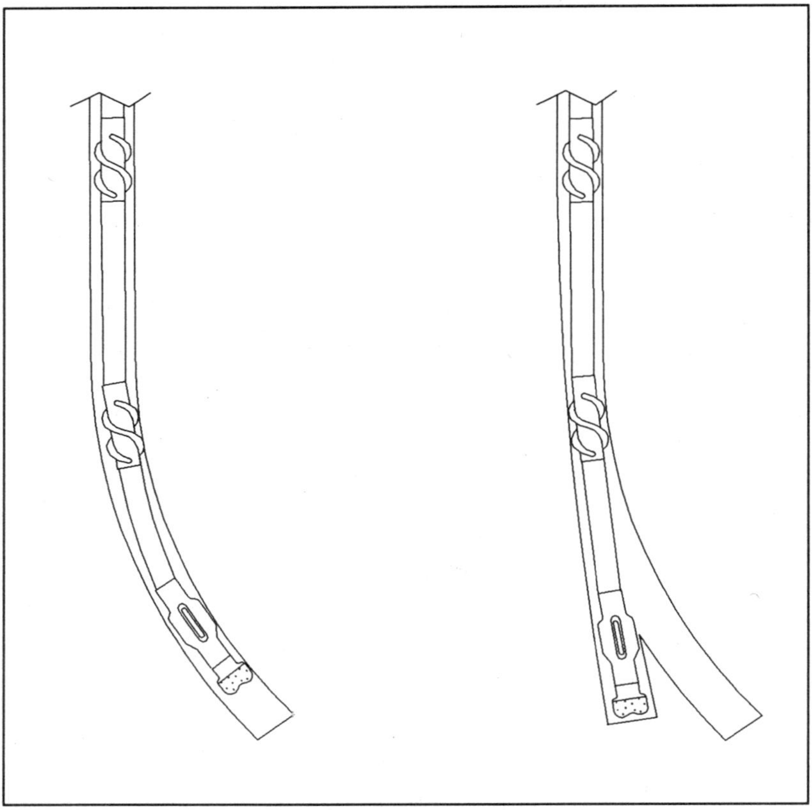

The hole changes direction abruptly at the point of sidetracking or deviation, with larger changes at higher rates of angle build. The curved section also may have rough walls. Drilling the straight, inclined hole section with a packed hole rotary assembly is the next normal directional operation. This assembly is large and stiff. It must bend and conform to the hole curvature as it passes into the curved angle-build section. Stabilizers and the bit on the bottom of the packed-hole assembly can hang on these rough places. If the assembly does not pass freely, it must be pulled and the buildup section reamed. Otherwise, the packed-hole assembly can stick or possibly part, causing a fishing situation.

Reaming does not reduce the general angle of build. It enlarges the curved section near the ends. This effectively reduces the angle a small amount, and smooths the walls of the wellbore so that other tools can pass freely.

It is important not to ream with the packed-hole assembly. It does not ream efficiently, and there is a high risk of accidentally sidetracking. The bit should not be used for reaming for the same reasons.

Sidetracking in the deviated hole almost invariably requires plugging back and sidetracking or deviating again. Sidetracking may be prevented in severe cases by replacing the bit with a stubby (2–4 ft) nose guide with a circulating port in the end. A very short nose guide may allow sidetracking and a long, slender nose may break off under severe reaming conditions.

The first step in reaming is to run a reaming assembly to the top of the curved section. Reaming operations must be conducted carefully to prevent accidental sidetracking. The hole is reamed by rotating and lowering the reaming assembly using minimum bit weight and a moderate rotary speed. It is important to keep the reamer moving vertically and to not ream for a long time at one point. Several moderately fast reaming passes are much better than one very slow pass. The reaming assembly may be too aggressive if the hole is very hard to ream. If so, it should be pulled and a less aggressive assembly should be run. After reaming, the next assembly must be run carefully because of the risk of sticking. If necessary, the section can be reamed again. Reaming with a string reamer is normally avoided in this situation.

Reaming can be a high-risk operation. Excluding accidently sidetracking, there are other risks. Mechanical failures include damage to tool joints on each end of the reamers and parting the drillstring by twisting off due to excess torque. These can cause a fishing situation. It is important to ream as smoothly as possible as a preventive action. The problems and risks of running tools into the angle-build section of the hole are important reasons to use the minimum buildup angle. A hole drilled with a buildup angle of 2°/100 ft seldom requires reaming, and then usually only a small amount. A build angle of 2.5°/100 ft often requires reaming and 3°/100 ft usually requires reaming. These assume passing standard assemblies. Horizontal patterns require higher build rates, adding to the risk of horizontal drilling. Reaming increases risks, causes problems, and increases costs.

STRAIGHT, INCLINED SECTION

The straight, inclined section is drilled with a hold rotary assembly. Sometimes very long, straight, inclined sections are drilled with a steerable motor assembly as described in Chapter 5. Packed-hole or stiff assemblies drill new hole with almost the same drift and direction as the previous hole. Formations often affect the drift and direction of the hole drilled with a stiff rotary assembly, even with the most rigid assembly, but usually the effect is small. For example, a formation may cause a certain limber assembly to build angle at a rate of 1°/100 ft. With a stiff assembly, this may be changed to 1°/400 ft, depending upon the stiffness and efficiency of the assembly. Therefore the stiff assembly will not drill an absolutely straight hole, but it is relatively straight for all practical purposes. This slight change of angle is normally allowed for by building the angle of the buildup section slightly higher.

The next step is to finish drilling the buildup section and pull the angle-build rotary assembly out of the hole. Normally, the hold rotary assembly is run when the drift angle is building but near (often about 90%) to the planned drift angle for the straight hole section. The stiff rotary assembly continues building angle at a decreasing rate. Then the angle stabilizes and the hole becomes straight after drilling 200–400 ft. A hold rotary assembly is connected to the drillstring and run into the hole, stopping about 100 ft above the kickoff point. The next step is to start rotating slowly and lower it carefully into the angle-build section. If the hole is tight, the assembly is pulled and the section reamed. It is not uncommon to encounter tight hole on the first tool run and sometimes on subsequent runs, even after reaming. It is important to always have a jar-bumper in the upper part of the stiff assembly for releasing it if it sticks.

The straight, inclined hole section is drilled with standard bit weight and rotary speed, monitoring drift and direction. Surveying starts at 30–60 ft intervals. The distance between measurements is increased to about 90 ft as drift and direction stabilize and assembly performance becomes more predictable. Measurement surveys should be included with a bit trip whenever possible to reduce measurement time. Regulatory agencies may require surveys at shorter intervals than needed for directional control. Sometimes it is possible to save time by taking the surveys at longer intervals while drilling and then run a wellbore survey later.

Trips should be made as necessary to change bits or equipment on the bottomhole assembly. Stabilizer wear reduces assembly rigidity. New stabilizers are full gauge, the same diameter as the bit, and the diameter reduces due to wear while drilling. They should be replaced when they are 1/8 in. undergauge (the diameter of the stabilizer is 1/8 in. less than the diameter of a new bit). It may be necessary to replace them when they are 1/16 in. undergauge if stiffness is critical. Bit walk often is a problem.

Stiff assemblies cannot be controlled efficiently in the horizontal direction. They are controlled in the vertical direction by adjusting the stabilizer placement and diameter and by the method of drilling. Correction runs should be made if necessary. Otherwise, the straight, inclined section is drilled into the target or objective formation to complete the bend-and-run pattern.

CORRECTION RUNS

Correction runs are procedures for changing the drift and direction of the hole. They are most commonly made while drilling with directional rotary assemblies, mainly to change the direction but also to change the drift angle if necessary. Horizontal direction cannot be controlled with a rotary assembly. If the hole direction changes, then it should be adjusted with a correction run. The same applies to the drift angle if the rotary assembly does not increase or drop angle correctly, but this situation is less common. Correction runs also are made for other reasons, such as changing the target location. Correction runs should be made whenever needed but avoided when possible because of increased costs. Any one of the three measurement systems may be used, depending upon the specific situation.

Correction runs are similar to deviating or sidetracking with a deviation motor assembly. The same procedures and precautions are used. The required changes in drift and direction are calculated similarly. The type of deviation equipment on the deviation assembly depends upon the changes needed. A common bent sub above the motor serves for normal drift and direction changes. A bent sub and bent-housing motor make larger changes. Orienting is more complicated for combined horizontal and vertical changes and normally is calculated with a computer program.

Next, the drillstring is pulled and a deviation motor assembly is run into the hole. Turn and direction are determined, orienting is done with these corrections, and directional drilling begins in the normal manner. Drift and direction are monitored while drilling and orienting is done again as needed. Assemblies should be lowered cautiously into the new curved hole section to prevent sticking after the correcting run. Reaming should be done if necessary. Drilling continues, correcting the drift and direction of the hole. Then the assembly is pulled out of the hole and operations resume with the most applicable drilling assembly.

DOUBLE-BEND

A double-bend or *S* pattern has two bends. It starts similarly to the bend-and-run pattern. After drilling the straight, inclined section, angle is dropped, and the second bend is drilled. The angle may be dropped by any amount, usually to 0° (vertical) and the hole is drilled into the target. Some patterns do not drop the angle of the hole to vertical but instead another inclined, straight section is drilled into the target (see Fig. 4–5).

Generally, angle-drop rotary assemblies are run in preference to angle-drop motor assemblies unless horizontal directional control is a problem. They are very efficient, and different modifications provide for dropping angle at various rates. There is less risk of sticking. They are strong, so there is a better chance of releasing them if they stick. The angle-drop motor assembly is not commonly used for dropping angle except in special situations. It is used in holes where formations have a strong tendency to change the course of the hole, either the drift angle or direction. Another use is for drilling patterns where drift angles require very precise control. Angle-drop motor assemblies are operated similarly to rotary versions.

The straight, curved buildup and inclined hole segments are drilled as described in the previous section about single-bend patterns. Then the stiff rotary assembly is pulled out of the hole. An

Figure 4–5
Drilling a double-bend directional well

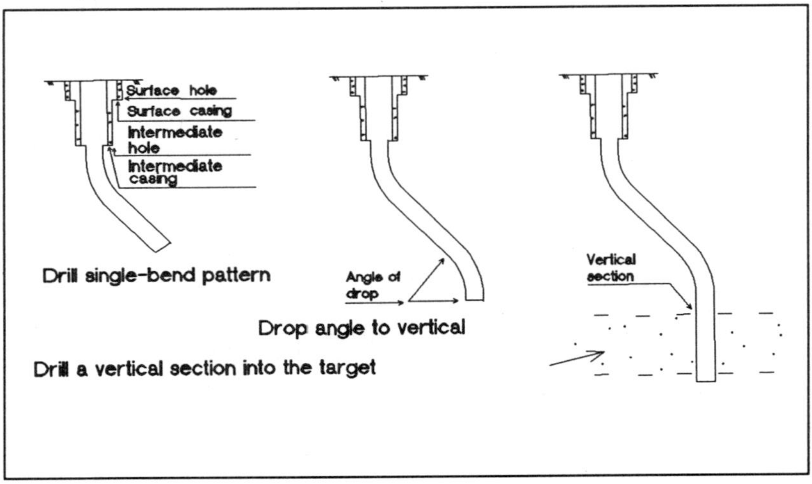

angle-drop rotary assembly (pendulum) is run into the hole. The angle-drop rotary assembly reduces the drift angle by drilling the hole in a curved downward direction. It should pass through curved hole sections without difficulty, but they can be reamed if necessary. Drilling begins in the normal manner, recording drift and direction measurements periodically. Correction runs can be made if necessary, but they are seldom needed. Trips should be made to replace worn bits or stabilizers or to change equipment on the bottomhole assembly.

The angle-drop rotary assembly drops angle faster at higher drift angles and at a reducing rate as the drift angle decreases. The angle can be dropped faster with slower rotation and reduced bit weight. The action may be reversed to drop angle at a slower rate. A packed-hole pendulum with two stabilizers is used for additional support at the fulcrum point in soft formations, out-of-gauge holes, and for similar conditions. The use of double stabilizers increases assembly efficiency in this case by distributing the lateral force over a larger area. Drag and torque increase with continued drilling, and there is a correspondingly higher risk of keyseating in the upper bend while drilling in and below the second bend. Wall sticking is more of a problem.

Drilling continues and the drift angle is reduced the required amount according to the pattern. A common design includes reducing the angle to 0° (vertical) with the angle-drop rotary assembly, excluding formation influence. Then drilling continues vertically into the target with the same assembly. The pendulum assembly also is a common assembly for drilling vertically in areas where the formations tend to cause crooked hole.

Another slightly more complex pattern includes drilling an inclined, straight section after reducing the angle a specified amount that is higher than 0°. More complex patterns, including horizontal turns with correction runs, are drilled by combining the different directional drilling procedures used for drilling single- and double-bend patterns. It is wise to use additional precautions when drilling these more complicated patterns because of the higher risks.

A special case of the bend-and-run pattern includes sidetracking a fish or junked hole and drilling vertically into the target formation. A lower buildup angle of about 10° normally is sufficient here. A sidetracking plug is set and then angle is built to about 10° by drilling 250 ft of vertical depth at an angle-build rate of 2°/100 ft. The new hole is a horizontal distance of about 40 ft from the old hole at this point. Then angle dropping begins with a pendulum assembly. It is dropped to vertical, and drilled vertically into the target. The vertical hole is at a horizontal distance of about 100 ft from the old hole, which is usually satisfactory. A higher angle of build is possible but seldom necessary. It also can create a dogleg situation causing keyseating and sticking while drilling the vertical hole deeper.

A modification is sidetracking blind. The first step is to sidetrack the old hole and drill the new hole as previously described, except that only drift measurements are recorded. Normally, directional measurements are unnecessary, especially for large targets. This procedure was common in the past, but the current practice is to measure both drift and direction.

EXTENDED-REACH

Extended-reach patterns have a long horizontal distance between the surface and bottomhole location. The basic design is similar to the bend-and-run pattern but with a longer inclined, straight section. It is possible to deviate or sidetrack the vertical hole, build angle, and then drill the straight, inclined section similarly to drilling a bend-and-run pattern. Drilling can be done with rotary or motor assemblies and measurements can be recorded with one of the three measurement systems, depending upon the pattern and hole conditions. The straight, inclined section may have a higher drift angle, sometimes approaching 80°, possibly increasing the difficulty of building angle. Drilling continues with the procedures described for the bend-and-run pattern for lower angles to about 60°. Higher-angled straight, inclined sections are drilled with the horizontal drilling procedures, using tangents if necessary, as described in Chapter 5.

Drilling problems for extended-reach patterns are similar to those in other directional drilling but more severe in deeper holes and at higher drift angles. Problems arise due to the increased drag and torque, and keeping the hole clear also can be a problem. Keyseats can develop in the curved kickoff and angle-build sections while drilling deeper. In severe cases, it may be necessary to run and cement casing through the buildup section and upper part of the straight, inclined hole to reduce or prevent the problem while drilling deeper. There is a high risk of wall sticking in the deeper sections of the inclined hole where the drillstring lies against the low side of the hole. High quality mud is a major help in alleviating these problems.

SLANT HOLE

Slant holes are drilled similarly to straight holes, but they start at an angle or slant from the surface and are drilled with slant-hole rigs. Slant holes tend to be a distinct category. They are characteristically shallow, seldom exceeding 4,000 ft TVD and 6,000 ft MD. The arbitrary definition of low- and high-angle is not directly applicable to slant holes. They approach the high-angle category by starting at an angle of 30°–45°.

Directional drilling procedures are similar to those for drilling a straight, inclined section or a single-bend pattern except that the hole starts at an angle. Most early slant holes were drilled with rotary assemblies, but either rotary or motor assemblies can be used. Measurements may be taken with any one of the three measurement systems depending upon the pattern and formation conditions.

The first step is to position the mast to point the drillstring in the correct horizontal direction toward the target. Then it is raised 30°–45° from the vertical. Drilling starts with a limber or modified angle-building assembly and continues for a short distance. This assembly is pulled and a packed-hole assembly is run, possibly modified to build angle slightly. This overcomes the slight angle-dropping tendency of the stiff assembly so that the resultant hole will be inclined but straight. Conductor or surface casing is run and cemented at shallow depths, usually several hundred feet. The procedure is similar to other directional casing operations, with allowances made for the angle.

The remaining straight, inclined hole is drilled into the target with a stiff, packed-hole assembly for a common, straight, or undeviated slant hole. Steerable assemblies are used in some cases. Trips should be made as necessary to replace dull bits or to change the bottomhole assembly. The stabilizer size and placement should be adjusted so that the assembly builds angle at a very low rate to maintain the drift angle in the straight, inclined hole section.

Slant holes are deviated for similar reasons as for other directional holes. Reasons include drilling to a new target, sidetracking to bypass a fish, and changing the direction of the hole. Any reasonable kickoff point can be selected below the conductor casing. A deviation assembly is run and oriented before deviation begins. Deviation and directional drilling procedures are similar to those described for single-bend or double-bend directional patterns. Holes with higher drift angles are drilled similarly to horizontal drilling as described in Chapter 5.

Most common drilling problems occur in slant holes. They generally are less severe because of shallow depths and lower drillstring weight. The drillstring may not run into the hole by gravity action due to the high angles. The main cause is the drillstring lying on the low side of the inclined hole and the frictional resistance to movement. The problem is more common while running the drillstring into the hole, unlike other directional drilling where the more severe dragging problems occur while pulling the tools out of the hole. Obtaining adequate bit weight also may be a problem. Slant-hole rigs have a pull down system, which literally pushes the drillstring into the hole. The pull down creates a downward force on the drillstring as needed during drilling and tripping and for running casing.

CASING AND CEMENTING

Casing is run and cemented in directional holes in the same general manner as for vertical holes. It is important to allow for problems imposed by the drift angle and operating in directional holes. It is almost analogous to running and cementing casing in very deep vertical holes with high casing loads. It is possible to run all standard casings, liners, stub liners, and tie-back liners. Some deviated wells use special applications of conductor and drive pipe as described in this chapter. Standard casing design principles are applied with provisions for higher loads due to drift angles and additional drag and torque.

Most regular and all critical casing and liner tubulars should be inspected with standard procedures. More detailed procedures for inspections should be used for deeper, complex patterns, especially with critical casing loads. The rig-hoisting equipment, including the drawworks breaking system, may require inspection for very heavy casing loads. Handling of heavier loads smoothly should be provided for by increasing the number of drilling lines between the traveling block and crown.

Liners should be set with either mechanical or hydraulic liner hangers, being careful not to overlook inspection of the hangers. Hydraulic hangers may be preferred in holes with higher angles with high drag and torque. It often is difficult to reciprocate and rotate for seating mechanical hangers under these conditions. It is important to make sure that the liner hanger has sufficient tensile strength with an adequate safety factor for longer and heavier liner loads, especially under conditions of high drag and torque. A clean mud system helps prevent the liner from plugging during running.

It is always important to rabbit the drill pipe or other pipe either immediately before or (preferably) simultaneously with running the liner.

Either automatic or manual-fill float equipment may be used depending upon conditions. Automatic fill equipment fills the casing automatically making it easier to run; many operators prefer it. It is important to be sure the float equipment does not plug. Cuttings suspended in the mud column pass through the automatic float equipment while the casing is being lowered into the hole. These cuttings tend to settle and accumulate partially because of the casing or liner movement. The cuttings also can plug the float equipment after circulation starts. There is a strong motivation to run the casing in a clean hole. It is necessary to ensure that the casing fills properly while being run into the hole with either type of float equipment. Otherwise, there is a risk of plugging or collapsing the casing.

Scratchers and centralizers should be designed for as needed; there is some preference for solid-body centralizers in holes with higher drift angles. The correct type and number of centralizers should be used, allowing for hole deviation. The next step is to determine the correct number, based on the standoff and calculated with computer programs. Proper casing-to-hole clearance must be ensured for good cement-to-formation bonding. The minimum, correct number of centralizers and scratchers should be used to reduce hanging up, dragging, and possibly sticking.

The cement slurry should be designed according to accepted practices. The cement is pretested as described in Chapter 3, designing for spacers and chemical washes as needed. The next step is to mix, pump, and displace the slurry and spacers in a workmanlike manner, including reciprocating and/or rotating casing and liners. Preference is given to batch mixing. The correct slurry density is obtained by observing recording densimeters and verified by weighing with a mud-weighting scale. It is necessary to catch wet and dry samples.

A good, high-quality mud in a clean hole will reduce many problems associated with running casing and liners and obtaining good cement-to-formation bonding. Two of the more common problems are high levels of drag and torque and the lack of good hole cleaning. These subjects are covered elsewhere, but they also may cause a failure during casing and cementing. Casing and liners are highly susceptible to differential pressure sticking during running and while cementing, even in holes with low drift angles. They are very difficult to release or recover by fishing and can cause a junked hole. High levels of drag and torque increase the risk of sticking. Free casing movement in a clean hole improves the quality of the cementing job, allowing better cement placement and bonding.

DRILLING PROBLEMS

Drilling problems are situations that restrict operations and increase risk and cost. Most problems are similar to those in vertical drilling but usually are more serious in directional and horizontal drilling. These range in severity from minor complications that are easily resolved to obstacles that cause the loss of the hole. Frequency and severity increase with increasing depth, higher angles, the number of angle changes, and the time spent on the operation. Most of these are preventable by planning, conducting prudent operations, and taking correct preventive actions. Excess drag and torque can be major problems in directional drilling and often are more severe in horizontal drilling as described in Chapter 5.

BLOWOUTS

Blowouts occur when formation fluids such as oil, gas, and saltwater, often under high pressure, flow into the wellbore and upward to the surface in an uncontrolled manner. Blowouts are the most severe problem encountered. There is a high risk of the loss of life and equipment. Most blowouts occur because of incorrect preventive procedures and equipment malfunctions. There are procedures and equipment for the early detection of blowouts. One of the main tools for this detects transition zones during drilling (see Fig. 4–6). There are equally good preventive procedures for controlling most blowouts before they become a severe problem.

Underground blowouts occur when formation fluids flow uncontrollably into the wellbore, travel along it for some distance, usually upward, and then flow into a lower pressure formation. Initially, this is not as serious as a blowout at the surface. Still, it can cause a loss of the hole and develop into a surface blowout. There is a risk of a blowout in almost all drilling operations. It is important to ALWAYS have good well control procedures, including a plan of action and frequent blowout drills.

PROBLEM FORMATIONS

Problem formations cause problems both during initial drilling and later while drilling deeper below the formations. High-pressure formations cause blowouts; those containing saltwater may

Figure 4–6
Transition zones detected by logging while drilling (LWD)
(courtesy of Halliburton)

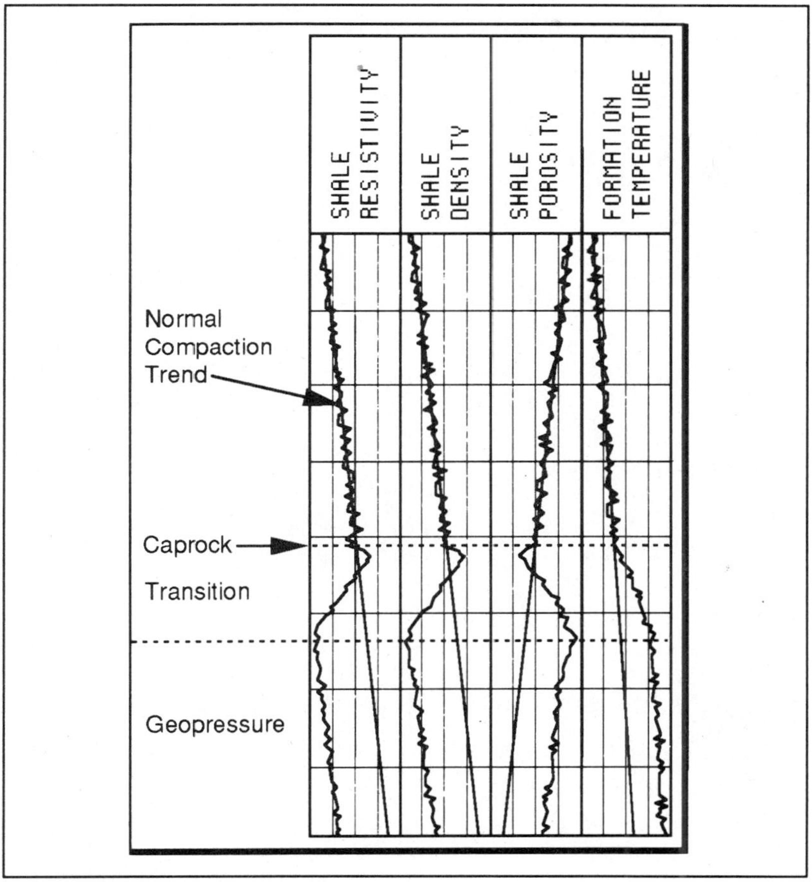

flow into the wellbore as a "saltwater flow." This contaminates the mud system, reduces hydrostatic head, and increases the risk of a blowout. Normally, the weight of the drilling fluid is increased. Lost circulation zones, or low-pressure formations, cause fluid loss. This restricts cuttings removal and increases the risk of differential pressure sticking. There is also a risk of a blowout situation if high-pressure zones are open in the wellbore. Lost circulation may be prevented by reducing mud weight, using plugging agents, sealing with cement, or running casing in extreme cases.

Crooked-hole formations affect the direction of the hole during drilling, sometimes strongly. They cause a crooked hole problem, increasing the risk of keyseats and wall sticking. They commonly dip steeply and may have laminations, often with alternating hardness. These should be drilled through with a straight hole, preferably vertical, when possible, for faster penetration rates. Hard or abrasive formations often severely reduce the penetration rate.

Fluid-sensitive formations such as bentonitic shales hydrate with most freshwater mud systems. They cause problems controlling mud properties and may slough into the hole. Sloughing and some fractured formations may fall into the wellbore, sticking drilling tools and causing fishing. Contaminating formations contain naturally occurring materials that contaminate many mud systems. The mud should be treated to minimize the effect or a resistant or inert mud such as oil mud should be used. Many of these formations cause out-of-gauge holes. This increases the difficulty of cleaning the hole properly to obtain good bonding when cementing casing. Most of these formations can be controlled with the correct type of high-quality mud.

Some formations contain hydrogen sulfide (H_2S), which is poisonous to humans and causes drill tools to fail prematurely. Hydrogen sulfide is extremely dangerous, even in very low concentrations. It is vital to always have special equipment and procedures for drilling in areas where the formations contain hydrogen sulfide. Other formations contain carbon dioxide (CO_2) or nitrogen, which are less dangerous but still cause problems.

CROOKED HOLE AND KEYSEATS

A crooked hole is a wellbore that has turns and bends that are commonly called doglegs. Crooked-hole formations, poor drilling practices, and deviation in directional drilling cause doglegs. Doglegs are measured as the degrees of change in the vertical angle per 100 ft of hole. Absolute dogleg is more accurate as a criterion. It includes the combined changes in both the horizontal and vertical directions. Doglegs cause drag, torque, and keyseats.

Keyseats are a slot or groove worn or cut into the side of the wellbore. The drillstring slides and rubs against the side of the wellbore during rotation and tripping, more in crooked-hole sections such as doglegs. Continued pipe movement wears the groove deeper, creating a keyseat (see Fig. 4–7). Smaller diameter tubulars, such as the drillpipe, slide through the keyseat. Larger diameter tools, such as the top of the drill collars (most common), bit and

Figure 4–7
Keyseats

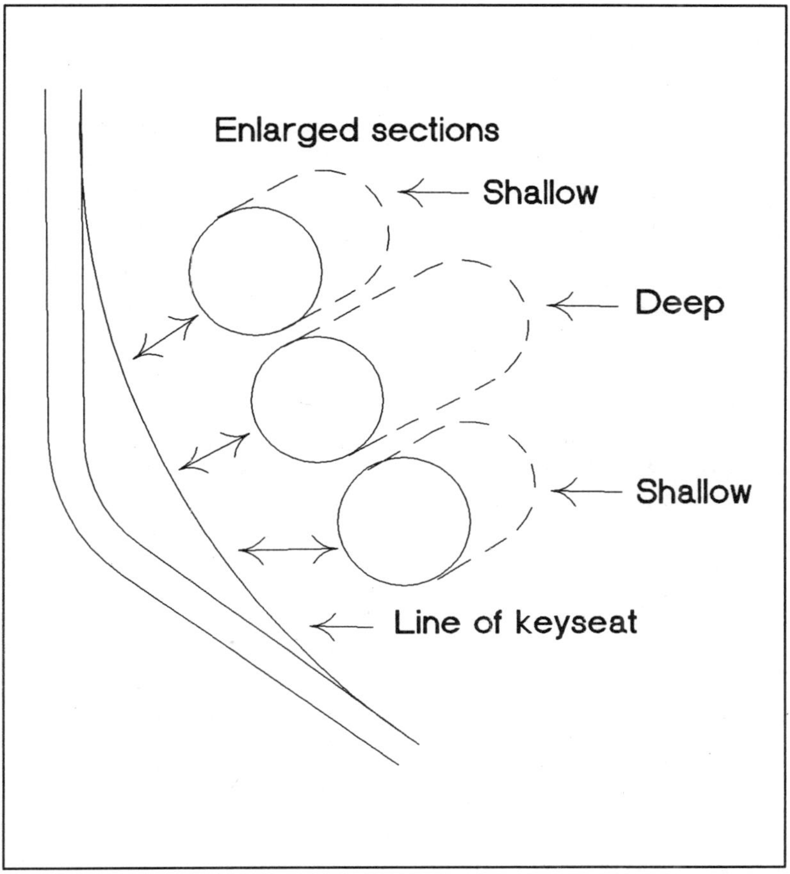

drillpipe tool joints (less common), cannot pass. They stick or keyseat, requiring special releasing procedures that are not always successful. This leads to sidetracking or redrilling the hole if they cannot be released or recovered by fishing.

The best method to eliminate keyseats is prevention. Smooth bends should be drilled through minimum changes of angle. Keyseats can be detected most of the time before they become a serious problem. Developing keyseats cause extra drag during trips, often periodic at 30 ft intervals as tool joints drag through the keyseat. It is important to watch for these signs and take the proper action. A keyseat wiper can be placed on top of the drill collars. Developing keyseats can be removed by reaming. It is best to design the drilling program to place casing through hole sections subject to severe keyseating.

Directional holes are intentionally deviated with one or more bends and turns. They require more time to drill and many trips for the drilltools. All contribute to developing keyseats, often a severe problem in directional and horizontal drilling. It is best to deviate the hole with a smooth curve within controllable limits whenever possible. The keyseats may be removed by reaming periodically. The risk of keyseating increases with higher angles as in horizontal drilling. The risk may be handled by recognizing the problem and taking all precautions as described in Chapter 5. Still, keyseats are a continuing, often major problem.

DIFFERENTIAL PRESSURE STICKING

Differential pressure or wall sticking occurs when the drillstring sticks against the side of the hole. It occurs opposite lower pressure formations under conditions in which the hydrostatic pressure exerted by the mud column is higher than the formation pressure. This creates a differential pressure into the formation, normally retained by the filter cake, by the invasion of mud-solids into the formation, and by impermeable formations. Wall sticking starts when a section of the drilltools, usually the drill collars, contacts the wall of the hole. Mud particles collect around the contact area between the collars and the wall of the hole, creating a sealed area. Differential pressure against this sealed area literally sticks the drill collars to the wall of the hole so that the drillstring cannot move.

Wall sticking occurs rapidly and without warning. This is too often overlooked or underestimated. Directional and especially horizontal drilling create almost perfect conditions for wall sticking. The pipe always lies against the side of the inclined hole. The assembly is at rest for connections, recording measurements, and for similar reasons.

It is possible to minimize the risk of wall sticking with various preventive procedures. Spiral drill collars reduce the surface area available for sticking. The force against the wall of the hole can be decreased by reducing the weight of the bottomhole assembly. A good quality mud with low weight and minimal solids can be used. It is important to keep the drillstring moving whenever possible, making connections rapidly and minimizing measurement time. Oil mud can be used in severe situations.

It is common to try releasing wall-stuck assemblies first by soaking with fluids that dissolve the wall cake causing the pressure seal. They can be displaced into the hole to the section where sticking occurs and allowed to rest or soak. This destroys the sealing mud cake, releasing the drillstring. Other releasing methods include reducing hydrostatic pressure displacing lighter fluids and using packers. Otherwise, the stuck drillstring may be recovered with fishing procedures.

HOLE CLEANING

Cleaning the hole is the procedure of removing drill cuttings and other formation particles from the wellbore. It is important to have clean drilling fluids in all holes. However, the effect is more important in directional holes and can be critical in horizontal holes. Holes that are not clean can cause many problems. The nozzles of jet bits may plug. Cuttings in the drilling fluid can plug casing float equipment and liner hangers and may affect the success of cementing the casing. These plugging actions shut down operations until corrected and can cause the drillstring to stick. Cuttings and small-sized solids retained in the mud system cause deteriorating mud properties that in turn cause other problems. Removing the solids from the mud system is more difficult, especially the finer-sized particles. Solid particles may settle and stack

Figure 4–8
Drill cuttings slump

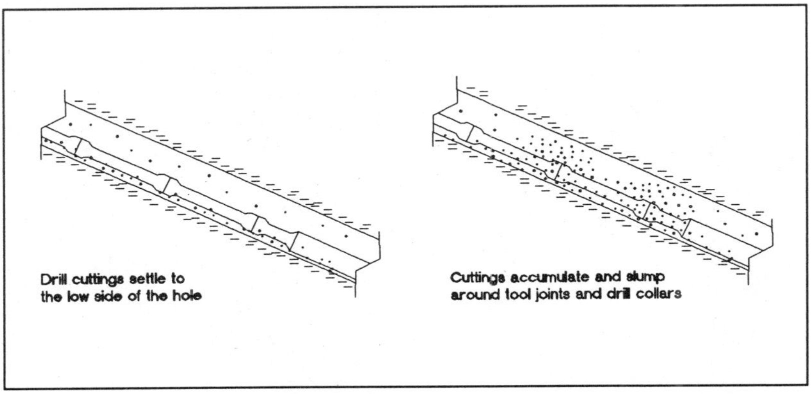

up or drag on the low side of the inclined wellbore. These accumulated solids in the wellbore may cause additional, often severe, problems. They slump or slide downward due to gravity (see Fig. 4–8). This can stick the drill tools, restrict circulation, and may cause lost circulation.

The hole cleaning problem is almost self-perpetuating, especially at higher drift angles. Increased drilling and circulating times decrease particle size. Smaller particles are harder to remove from the mud system and may be recirculated into the hole. This, combined with additional circulation, erodes and enlarges the wellbore. Mud moves through enlarged sections at a slower rate, so more cuttings settle and accumulate.

Normally, cuttings and smaller solid particles are removed by circulating drilling fluid. But it is more difficult to remove cuttings from an inclined hole than from a vertical hole, and it is more difficult yet for horizontal holes. The solid particles roll and drag on the low side of the hole as they move upward with the moving mud column. This action reduces particle size. Drill tools rotate and move against the low side of the hole, grinding and crushing the particles, which further reduces particle size. Normal fluid flow in the drillpipe section of the hole is laminar with less fluid movement near the wall of the hole. This in turn reduces the upward movement of the solid particles. Combinations of these actions retard hole cleaning.

In summary, holes with higher drift angles require correspondingly longer circulating times, especially for horizontal holes. It is estimated that up to 3 times as much circulating fluid volume is needed for cleaning a hole inclined at 45° compared to a vertical hole, and even larger volumes are needed at higher drift angles.

Selecting the correct drilling fluid is one of the most important remedial actions. It must have good flow properties, suspend all particles, and transport them to the surface. Cuttings may be flushed out of enlarged hole sections by circulating small (20–50 bbls) volumes of a viscous fluid. Higher mud density may help in extreme cases. Equivalent circulating density (ECD) helps in evaluating the circulating conditions in the hole. ECD provides a measure of the pressure loss in the annulus between the bit and surface, normally expressed in pounds per gallon (PPG). Higher values generally suggest deteriorating mud and circulating conditions. As many of the solid particles as possible should be removed at the surface to prevent circulating them again and compounding the problem. Good circulation and large capacity, efficient solids-removal equipment are very important and cannot be overemphasized.

TUBULAR AND EQUIPMENT WEAR

Drill tools, equipment, and casing are subject to wear during drilling operations. Lowering, lifting, and rotating the drill tools while drilling cause wear. Sand and other solids in the mud system also cause and accelerate wear. Severity increases with depth and at higher angles in directional and horizontal holes, especially at bends and turns. Wear causes leaks and equipment failures leading to fishing.

Wear cannot be eliminated but it can be kept to a minimum. Actions that minimize wear include good well planning and the use of correct equipment. Wear may be reduced with a clean mud system, a low-weight drillstring, and speeding up operations so that there is less pipe movement in the hole. A bar magnet in the flowline may show metal particles, indicating tubular wear. It is prudent to inspect drill tools periodically and replace worn equipment.

Casing wear should be provided for in planning and design by using heavier weight casing when required, especially through hole sections with high-angle bends and turns. Casing caliper and other logs show worn casing. The first step is to run a base casing inspection log shortly after running casing in critical wells. Then inspection logs are run at periodic intervals to evaluate the condition of the casing. Drillpipe rubbers help to prevent drillpipe and casing wear. Rubbered drillpipe can be used inside casing with double rubbers in bends and high-angle sections. Many mud additives for increasing lubricity are of questionable value. About 6 percent diesel oil in water-base mud improves lubricity. Oil mud has good lubricity and minimizes wear.

DRILL TOOL MAINTENANCE

The drillstring is a long string of drillpipe, drill collars, and other tools. As with all equipment, good maintenance is mandatory and prevents many problems. Good maintenance starts with the design of the well plan and drilling assemblies. There can be hundreds of tool joint connections in the drillstring. A failure at any one can cause a fishing job. Tool joints must be in good condition and tightened correctly to the recommended torque. They are subject to fatigue failure that increases with rotational speed and larger changes of angle. There are long periods of circulation, frequently at high and fluctuating pressures. Connection leaks are very difficult to detect under these conditions before they cause a failure and result in a fishing situation.

All drill tools should be inspected periodically and replaced or repaired as required. New equipment should be inspected when adding it to the assembly. The use of ultraviolet (black) light is the most common inspection procedure for tubulars, and sonic inspection is sometimes used for other equipment. Frequent visual inspections must be made. It is a good practice to change the order of stands on trips. This prevents running the same joints of drillpipe in deviated hole sections for an extended period. It is also helpful to alternate the order of break for stands of drillpipe on trips. As always, all tools and equipment must be operated within design specifications.

FISHING

Fishing is the process of recovering or otherwise removing a fish so that normal operations can be resumed. The fish is any obstruction or equipment (usually tubular) left in the wellbore that restricts operations. Fishing often is a series of complex, detailed, high-risk operations. Risk increases with the complexity of the fishing job and increasing depth. Fishing often is more difficult in directional holes than in vertical holes, but fishing operations are similar. Special and often severe fishing problems occur in high-angle and horizontal holes as discussed in Chapter 5 (see Fig. 4–9).

All fishing jobs are not successful. Fish are bypassed at greater depths if they cannot be recovered. This alternative is also considered when the cost of fishing becomes excessive or approaches the cost of sidetracking. Sidetracking may not be practical in some cases because of conditions such as casing and hole size requirements. Severe casing damage causes a junked hole, so the best procedure may be plugging and redrilling.

Fish may be recovered with various procedures using fishing tools and other equipment. A fishing assembly is similar to a short limber assembly with fishing jars and a bumper sub below a few collars. A fishing catch tool connected to the bottom of the assembly catches hold of the fish. Fishing jars and a bumper sub deliver sharp, heavy blows in upward or downward directions to release stuck fish for recovery. A milling assembly has a mill on bottom instead of a catch tool to remove small fish by milling. Stuck pipe logs which are run inside the stuck drillstring show where and how strongly the tools are stuck. Freepoint logs locate the point where the fish is free.

Figure 4–9
Fishing

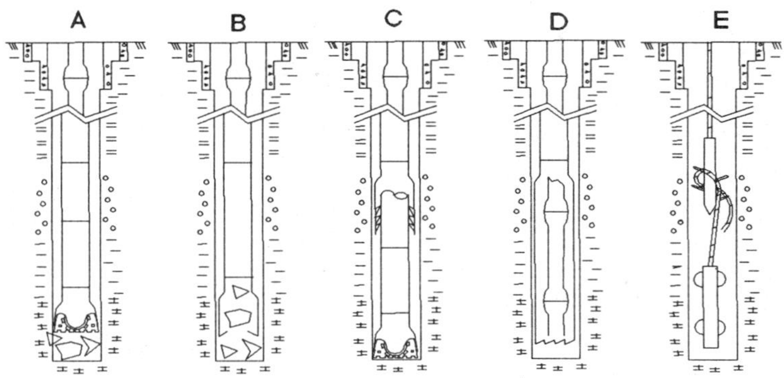

A = Drill up or "wall off" small junk
B = Recover junk in a junk basket
C = Catch a fish with an overshot
D = Wash over a fish with washpipe
E = Recover wireline fish with a wireline spear

The most common fish is comprised of one or more bit cones. It is possible to remove them by drilling with special bits and mills, which allow them to be recovered with junk baskets. Larger objects, such as bits, are broken into smaller pieces with a jet explosive or junk shot and then the pieces are recovered similarly to recovering bit cones. When the drillstring sticks, it can be worked by reciprocating, rotating and circulating, and perforating it if applicable. Wall sticking may be treated as described previously. If the drillstring (or sometimes the casing) is not released in this manner, then the stuck point is located with freepoint or stuck-pipe logs. The drillstring is parted above this point by backing-off or cutting and recovering the free section. Then the remaining stuck fish may be retrieved by one of several methods.

Larger fish, such as drillpipe and drill collars, are most easily recovered by screwing back into the top connection, working them free, and pulling them out of the hole. It is possible to run an overshot over the top of the fish and catch it on the outside if the top thread connection is damaged. A drillpipe spear enters a smaller fish, such as drillpipe, to catch it on the inside. New, clean tops are milled on fish with damaged tops, such as twisted-off drillpipe. Then the new top is caught with an overshot or spear and the fish is pulled out of the hole.

It is possible to wash and drill over a tightly stuck fish with large diameter wash pipe to release it for recovery. A very long fish is caught, parted by backing-off, and recovered in two or more sections. Large diameter fish are caught and recovered with a casing spear. A hole in the casing (such as a worn place) may be closed with a casing patch or by squeezing with cement. The parted casing sections are reconnected with a casing bowl. Wireline fish are caught and recovered with grabs or wireline spears.

Almost all fishing situations can be prevented. This requires planning, selecting the correct equipment, operating within design limits, and conducting all operations in a careful manner.

BIBLIOGRAPHY

D. D. Baldwin, R. W. Royal, and H. S. Gill. *Drilling High-Angle Directional Wells.* PD5(2). 11th World Petroleum Congress. London, 1983.

W. B. Bradley, et al. "Task Force Reduces Stuck-pipe Costs." *Oil & Gas Journal* (May 27, 1991): 84–85.

R. J. Crook, S. R. Keller, and M. A. Wilson. "Deviated-Wellbore Cementing; Part 2—Solutions." *Journal of Petroleum Technology* (August 1987): 961–966.

J. M. Davis and K. T. Corbette. "Polishing Drill Pipe Protectors Speeds Torque Reduction." *Petroleum Engineer International* (August 1991): 48–53.

R. D. Edwards and G. Strelkov. "Slant-Hole Drilling Finds Expanding Role in Canada." *Petroleum Engineer International* (February 1988): 20–26.

F. Harvey. "Horizontal Wells 4—Fluid Program Built Around Hole Cleaning, Protecting Formation." *Oil & Gas Journal* (November 5, 1990): 37–41.

R. C. Haut and R. J. Crook. "Primary Cementing: Optimizing for Maximum Displacement." *World Oil* (November 1980): 105–106.

T. Hemphill. "Tests Determine Oil-Mud Properties to Watch in High-Angle Wells." *Oil & Gas Journal* (November 26, 1990): 64–70.

P. Herbert. *Drilling with New-Generation Positive Displacement Motors.* SPE 10239. Society of Petroleum Engineers. San Antonio, TX, October 5–7, 1981.

W. Jones. "Horizontal Wells 3—Unusual Stresses Require Attention to Bit Selection." *Oil & Gas Journal* (October 22, 1990): 81–85

S. R. Keller, et al. "Deviated-Wellbore Cementing; Part 1—Problems." *Journal of Petroleum Technology* (August 1987): 955–960.

G. Kempt. *Oilwell Fishing Operations: Tools and Techniques.* Houston, Texas: Gulf Publishing Company, Book Division, 1986.

W. King. "Selecting Bits for Extended Reach and Horizontal Wells." *World Oil* (April 1990): 53–60.

M. Lesage, et al. "Pore-Pressure and Fracture-Gradient Predictions." *Journal of Petroleum Technology* (June 1991): 652–654.

M. Lesage, I. G. Falconer, and C. J. Wick. "Evaluating Drilling Practice in Deviated Wells with Torque and Weight Data." *SPE Drilling Engineering* (September 1988): 248–252.

J. D. A. McKee, T. Geehan, and B. Smolen. "Efficient Solids Control Key to Incentive Drilling Performance." *Petroleum Engineer International* (April 1990): 38–48.

K. K. Millheim and M. C. Apostal. "The Effect of Bottomhole Assembly Dynamics on the Trajectory of a Bit." *Transactions of the American Institute of Mining and Metallurgical Engineers* 271 (1981): 2323.

Ocean Industry. "Conoco Drills 17,800 ft with One Bit." *Ocean Industry.* (December 1984): 45.

D. P. Salisbury and C. K. Deem. "Tests Show How Oil Muds Increase Shale Stability." *World Oil* (October 1990): 57–65.

M. H. Seeberger, R. W. Matlock, and P. M. Hanson. *Oil Muds in Large Diameter, Highly Deviated Wells: Solving the Cuttings Removal Problem.* SPE/IADC 18635. Society of Petroleum Engineers. New Orleans, LA, February 28–March 3, 1989.

J. A. Short. *Fishing and Casing Repair.* Tulsa, Oklahoma: PennWell Publishing Company, 1981.

J. Smith and B. Edwards. "Slant Rigs Offer Big Payoffs in Shallow Drilling." *Oil & Gas Journal* (March 30, 1992): 64–66.

P. H. Tomren, A. W. Iyoho, and J. J. Azar. "Experimental Study of Cuttings Transport in Directional Wells." *SPE Drilling Engineering* (February 1986).

M. Zamora and P. Hanson. "Rules of Thumb to Improve High-Angle Hole Cleaning." *Petroleum Engineer International* (January 1991): 44–51; and (February 1991): 22–27.

CHAPTER 5

HORIZONTAL DRILLING

SUMMARY

Horizontal wells are drilled through curved sections up to a 90° angle and then horizontally into the formation. The three pattern classifications are short, medium, and long turn radius patterns. Short-turn patterns are drilled from cased wells with whipstocks and articulated pipe. Medium-turn patterns are drilled in larger diameter cased holes with slim-hole techniques. Otherwise, both medium- and long-turn patterns are drilled in open holes. Mostly motor assemblies and some rotary assemblies are used, depending upon the drilling situation. Tangents help to place horizontal sections correctly in the formation. Extended-reach and combination patterns are drilled by various, similar techniques. Formations should be evaluated by special well logging procedures and the data recorded with some measurement-while-drilling instruments. Casing or liners are run and cemented with a high-quality slurry. Isolation is improved with inflatable packers. The well is completed by standard perforating and stimulation techniques. Predrilled or slotted, uncemented liners are used for some open hole completions. Some wells flow naturally and others use artificial lift, such as pumping. Horizontal drilling is a complex, high-risk operation. Major problems include controlling direction, high angle-build rates, operating through curved sections, high levels of drag and torque, and thorough hole cleaning.

OPERATIONS

Horizontal and high-angle drilling operations generally are similar to directional drilling but more complex because of higher build rates and drift angles, and tangent and horizontal sections. The discussion referring to horizontal drilling generally applies to high-angle extended-reach patterns unless otherwise noted. Horizontal and extended-reach drilling described here includes angles greater than about 60°, more commonly about 70°–90°. There are indications that drilling straight sections with drift angles of 70°–90° are similar. Holes with low angles of 60° or less are described in Chapter 4.

It is possible to plug back and sidetrack medium- and long-turn holes in either the curved or horizontal sections. But the procedure should be used sparingly because it increases the difficulty of drilling a pattern that often is already complex. Mud logging equipment is run on most wells to aid in drilling, support hole guidance, and help in formation evaluation.

Most drilling problems found in other forms of drilling occur in horizontal drilling operations. The major problems encountered in directional drilling as described in Chapter 4 also occur in high-angle and horizontal drilling, often more frequently and with a higher degree of severity. Problems increase with increasing depth, higher angles, and longer horizontal sections. A few of these are summarized for emphasis and special applications to horizontal drilling.

In horizontal drilling, high stresses in equipment and tubulars are common. Good hole cleaning often is difficult to attain, but a clean hole solves many problems. Fishing is difficult and less successful as described at the end of Chapter 4. These problems cause high risks in horizontal drilling operations and emphasize the importance of planning and prudent operations. There were early concerns regarding horizontal holes remaining open. Hole closure by caving formations was not a major problem in early slant-hole drilling and later extended-reach drilling. It occurs in horizontal drilling, but it is not a severe problem.

DRILLING GUIDES

Drilling guides are special measures applicable to horizontal and other high-angle holes. These patterns are drilled with standard land and marine drilling rigs using standard drilling equipment with a depth rating approximately equal to the measured depth of the horizontal hole with a 10–20% safety factor. Top drives can improve drilling efficiency with steerable assemblies and help to handle difficult drilling conditions such as tight hole problems. It is important to select a high quality drilling fluid with good physical and chemical properties. Liquid drilling fluids should be used; a few holes have been air drilled with special measurement-while-drilling instruments. Adequate size of pumps, mud handling, and solids separation equipment must be ensured. This is very important and cannot be overemphasized.

Measurements in the upper part of the curved hole section are recorded with one of the three common measurement systems, although the magnetic single-shot is less common. Measurement-while-drilling (MWD) is more efficient and most commonly used in higher angle and horizontal hole sections. Accurate measurements are always important, especially since instrument errors tend to increase at higher angles.

Measurement tool systems and instruments should be evaluated carefully concerning their individual advantages and disadvantages before a selection is made. Various suppliers offer steering tools and a larger number have MWD systems. Instruments from each supplier may measure and record data differently and have varied capabilities and limitations, especially the more commonly used MWD systems. The measurement system and individual instrument(s) in the system should be selected to best serve the requirements of the specific project under consideration. MWD is more efficient for many horizontal and high-angle applications. Some MWD systems record lithology and other data that is very helpful for drilling and positioning the horizontal lateral correctly. It is important to use the correct length of nonmagnetic drill collars. Magnetic tools such as steel stabilizers should not be placed between nonmagnetic collars.

Most horizontal holes are drilled with motor assemblies. They build angle at higher rates and provide good directional control while building angle and drilling holes with higher drift angles. Rotary assemblies are used less often because of low angle-build rates and lack of horizontal directional control. It is common to drill with steerable motor assemblies as often as possible because of good directional control. They also serve for drilling in two modes as described in Chapter 2. It is possible to use either procedure, alternating periodically as necessary, depending upon the amount of directional control required. This is a distinct advantage, often saving tripping the drillstring to change the bottomhole assembly such as to install an assembly with a less aggressive climb rate. The action of the steerable motor assembly may be simulated with a regular motor assembly by drilling side-to-side as described in Chapter 3.

Guidelines for bit selection are similar in both horizontal and directional drilling. Greater preference is given to premium grade bits and shorter shanks with reinforced side-cutting action for improved directional control while drilling higher angle curved sections.

Drilling jar-bumper subs are an important and integral part of all drilling assemblies. They are always used except in specialized conditions. Sometimes it is necessary in high-angle and horizontal holes to divide or split assemblies into two parts. A jar-bumper sub placed near the top of the upper assembly effectively aids in releasing the stuck upper assembly and some length of pipe below it. However, the jarring action is less effective for releasing sections of the drillstring stuck a long distance (300–400 ft or more) below this upper jar-bumper. The single drilling jar-bumper is not sufficient for releasing the lower half of the assembly if it sticks.

Two (or double-drilling) jar-bumper subs are used sometimes when running split assemblies. A drilling jar-bumper sub is placed in the upper part of the drill collar assembly, is set to trip, and jarring begins at a normal level of overpull. A second drilling jar-bumper sub is placed near the top of the second or lower part of the assembly. It is set to trip and jarring begins at a lower level of overpull compared to the upper set. The jarring force is less compared to that needed by a jar bumper on the top of the upper part of the assembly set for tripping at a higher force. Still, it has a better chance of releasing a stuck lower section. The drilling jar-bumper is near the lower assembly, and the resulting jarring action is closer to the stuck point. The jar bumper must be placed below the required number of drill collars, usually three or four, based upon the specifications of the jar bumper. These collars supply weight for the jarring blow. Double-drilling jar-bumper subs improve the chance of releasing stuck tools.

STRESSES IN TUBULARS

Basic drilling ideas common in vertical and directional drilling require modification in some horizontal drilling patterns. Both directional and horizontal drilling patterns may have to be modified, especially to drill medium-turn holes with high-angle curved sections. One major change is the idea of operating part of the drillstring in compression. Drillpipe operates in tension in most vertical and directional drilling; otherwise, there is a high risk of parting and a fishing situation. Conventionally, bottomhole assemblies (BHA) operate partially in compression for applying weight to the bit. Larger, heavier drill collars with heavy-duty connectors withstand the severe forces caused by operating in compression. This is one main reason for their use as previously described in the section about free point.

BHA's operate in a similar manner in horizontal drilling. But, compression pipe and sometimes drillpipe may necessarily operate in compression in some high-angle and horizontal holes. The tubulars may be curved to a high degree and subjected to bending and buckling stresses. Bending stresses in curved tubulars cause tensional forces in the wall of the pipe on the outside bend and compression forces in the wall on the inside bend (see Fig. 5–1). These forces alternate rapidly during rotation and by some horizontal movements, subjecting the tubular to failure due to fatigue and embrittlement.

Drillpipe has about the same strength in compression as in tension if supported as a fixed column so that it cannot move in the lateral direction. This is approximately correct for pipe in tension in normal drilling. However, it can bend and possibly buckle when in compression. Several factors in horizontal and high-angle drill-

Figure 5–1
Bending stresses in curved tubulars

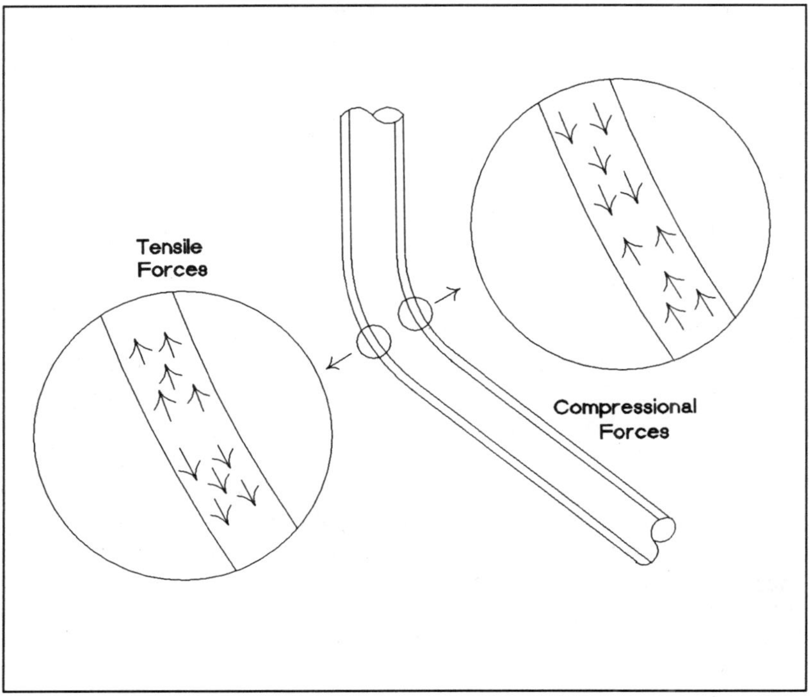

ing favorably affect operating the drill tools in compression. Rotating the bit with a motor or turbine reduces the torque on the drillstring a little. The drillpipe lies on the low side of the inclined or horizontal hole, so there is some lateral support to help prevent bending and buckling actions. This seems to increase as inclination increases. At high inclinations, drillpipe can withstand substantial compressive forces as indicated by calculations and confirmed by experience. There is less risk of failure in drillpipe connectors because they are stronger than the pipe body. It is important not to overdesign, but provisions should be made for higher tension, torque, and special stress situations.

All directional and horizontal holes have bends and turns, so dogleg, and more importantly absolute dogleg, are a natural result. Doglegs cause drag and torque and keyseats as described in Chapter 2. They also cause bending stresses and resulting tubular failures. Permissible doglegs have acceptably low angles, so there is minimal risk of damage to the drilling tools. Still, there may be a risk of keyseating and wall sticking, even at these low angles.

Drillpipe is susceptible to fatigue failure while drilling below a dogleg due to bending and flexure stresses. The amount of permissible dogleg depends upon the size and weight of the drillpipe, the weight suspended below the dogleg, and the rotational speed. String reamers are at risk of failure at the tool joints because of a similar bending action. Bending stresses are cumulative over time, an important reason to investigate the operating history of tubulars, especially drillpipe. It is possible to calculate absolute doglegs and more conveniently locate them on charts based on changes in vertical and horizontal angles. Permissible dogleg may be found using empirical monographs.

EXCESS DRAG AND TORQUE

Drag is a force restricting the movement of the drill tools in directions parallel to the well path. Torque is the force resisting rotational movement. Drillstrings rub and slide against the walls of the hole during rotation and tripping as part of regular drilling activities. Drag and torque are measurements of this frictional resistance to the movement of the drill tools. They occur in all holes. Drag is measured in thousands of pounds over or under the free hanging weight of the drillstring. Torque is measured in foot-pounds of applied torque. It is important to have a good weight indicator and torque-measuring equipment.

Drag and torque increase with an increasing number of bends and turns and higher drift angles. Drag and torque caused by deeper bends and turns can be highly amplified in shallower bends and turns. The deeper bends and turns cause a level of drag and torque on the drillstring. This causes a lateral force in the drillstring at shallower bends and turns. This can increase the drag and torque in the drillstring at the surface to a much higher level. The action is analogous to the use of a cathead where one wrap gives a certain level of pull, but two wraps can give a pull which is an order of magnitude greater. Other conditions that increase drag and torque include irregular wellbore walls, larger drill tool diameter relative to the diameter of the hole, thick mudcake and high-gel drilling fluid. Drag and torque are higher in open holes than in cased holes. Tool joints, stabilizers, and other projections on the drillstring tend to dig into the walls of open holes creating a dragging, plowing effect that further increases drag and torque. The dragging and wearing effect is more severe at bends and turns, frequently causing keyseats and related problems.

Excess drag and torque cause directional drilling problems, often very severe in horizontal holes. The drillstring can part from tension due to excess drag or twist off due to excess torque. Either case leaves an obstruction in the hole requiring fishing. Open hole drag causes keyseats that in turn increase drag and torque. Drag increases the risk of sticking in keyseats and differential pressure sticking. Drag also reduces available bit weight, severely at higher angles.

Eliminating all drag and torque is not practical, but preventive actions help reduce it to acceptable levels. It is best to design the well pattern for a minimum number of changes of angle and a low angle of build or drop. Excess drag and torque are reduced by placing casing in the hole. Drag and torque still occur, but casing eliminates problems of keyseating, differential pressure sticking, and the plowing effect. Drillpipe rubbers reduce casing wear, and double rubbers are run in bends and turns. Drillpipe rubbers should not be used in open holes because the rubbers wear excessively, and there is a risk of loose rubbers sticking the drillstring.

Reaming reduces drag and torque caused by keyseats and rough, uneven wellbores. It is important to drill smooth curves and straight "straight, inclined" sections. Drag and torque increase with increasing drillstring weight, such as occurs when drilling the hole deeper. Reducing drillstring weight reduces drag and torque. Weight reductions are increasingly effective at greater depths, such as for the bottomhole assembly in horizontal sections. Split bottomhole assemblies can be very effective. Tapered drillstrings may be helpful. Aluminum drillpipe reduces weight but causes operating problems.

High quality mud with good chemical and physical properties is essential. Oil mud should be considered for more demanding situations because of its good lubricating qualities. Water-base mud lubricity may be increased by adding 4–7% diesel oil with emulsifying agents, and mixing it thoroughly. Other lubricants such as asphalt, alcohol-base lubricants, and graphite give questionable results. Granular material such as walnut hulls are very effective in directional holes if used correctly. Walnut hulls may be used because of reduced cost compared to plastic and glass beads. They are equally effective if applied correctly.

FISHING

High-angle and horizontal holes present special problems that prevent some of the more useful fishing procedures. Many fishing procedures use tools lowered into the hole on wirelines or with shielded electrical conduits, which are commonly called wireline tools. Wireline tools move downward by gravity action and are retrieved by the cable. They cannot be run through hole sections with angles greater than about 60° in the conventional manner. The drag of the tool and wireline on the side of the wellbore overcomes the force of gravity, and the tool stops. Wireline tools sometimes may be run on coiled tubing or small pipe, but most of these have depth limitations. The tools may be pumped down with a plunger arrangement on top of the wireline tool and a pack-off on top of the drill pipe, which is similar to running logging tools in high-angle and horizontal holes. These unconventional methods of running wireline tools apply to a few cases but in general have limited applications in most fishing procedures.

A plugged drillstring is a very common fishing situation. Bit plugging is moderately common during drilling and may occur after sticking the drillstring. It is important to resume circulation as quickly as possible because circulating mud helps prevent sticking (or additional sticking if the drillstring is already stuck). Circulation often is a strong measure to prevent or control blowouts. It may be possible to pull the drillstring out of the hole, depending upon specific conditions, but frequently this is not an option. The difficulty of cleaning solid particles out of high-angle and horizontal holes contributes to the problem. THE HOLE MUST BE KEPT CLEAN.

It is common to establish circulation by perforating the drillpipe or drill collars immediately above the point of plugging. Plugged bits are opened by blowing the jets out with an explosive charge lowered into the hole on a wireline. These procedures must be done rapidly under normal conditions, but this option is not available in high-angle and horizontal hole sections because the wireline tools will not fall as noted. Also, wireline tools cannot be pumped down against a plugged bit. Using coiled tubing here is very questionable, even if possible, because of the risk of leaving additional junk in the hole and further complicating the fishing operation. One of the few remaining alternatives in this situation is parting with a blind back-off, followed by washing over to recover the fish. This is difficult under the best circumstances, even excluding the additional limitations imposed by high drag and torque and the high-angle or horizontal hole.

If the stuck drillstring can be circulated, then wireline tools may be used in a limited fashion. But fishing options are very limited without circulation. Freepoint and stuck-pipe logs cannot be run to find the section of sticking. The drillstring cannot be perforated to establish emergency circulation. It cannot be parted with a chemical or jet cutter or conventionally by backing-off with a string shot. Plugged drillstrings create very serious situations in high-angle and horizontal holes.

SHORT-TURN

Short-turn horizontal holes have a turn radius of a few feet to 60 ft and angular build rates of 95°/100 ft to greater than 1,000°/100 ft MD. Horizontal section lengths range from about 100 to a maximum of about 800 ft in a few cases. It is common to drill the pattern in cased holes with smaller diameters. Multiple, small-diameter horizontal holes are drilled extending radially from the same wellbore with some systems, but usually not with more than two holes. Short-turn horizontal holes generally are different from other horizontal classifications. They have special whipstock deviating systems and do not use conventional tubulars except possibly some very small sizes in and below the curved section. These patterns are not as common as other horizontal classifications.

Short-turn equipment and procedures can be complicated. It is small-diameter equipment, so it is weaker and more likely to fail. Drilling rates may be very restricted in harder formations. There may be problems with directional control while drilling the lateral with some systems. Normally, conventional tubulars cannot be run through the curved section (see Fig. 5–2).

Figure 5–2
Drilling short-turn horizontal wells

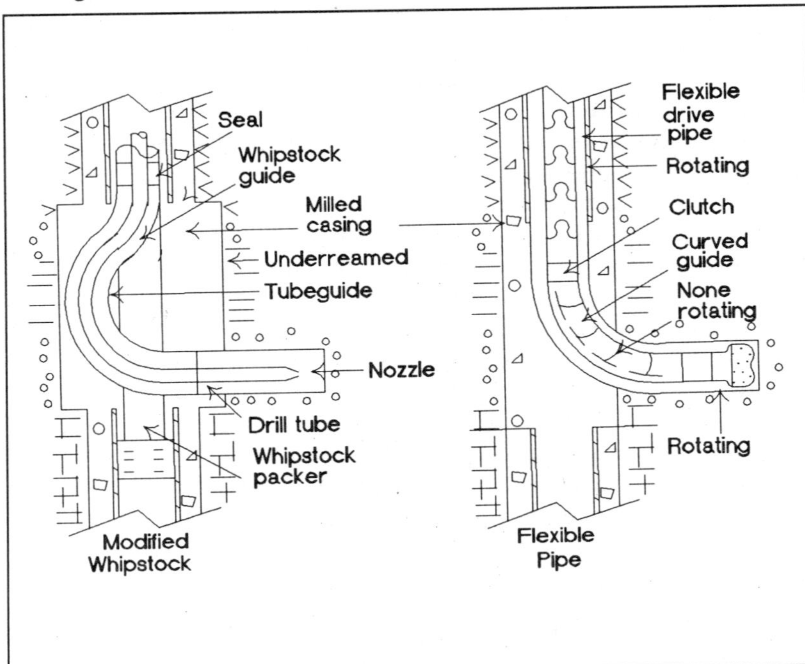

VERY SHORT-TURN

Very short-turn horizontal holes, sometimes called drainholes, turn the hole from vertical to horizontal in a few feet. They are drilled in previously cased holes, and often multiple horizontal laterals are drilled from the same wellbore. Sidetracking is done with a whipstock deviating tool with a curved guide. A long, slender steel tube fits inside the drillpipe and into the top of the whipstock guide. The upper end of the tube has a pressure seal to contain pressure and divert drilling fluid through the tube. A jet nozzle fits on the lower end of the tube. The tube is retrieved, and the drilling rate is controlled with a retaining cable connected to the top of the tube.

The hole is prepared first by plugging the lower hole as required and removing a section of casing by milling as described in Chapter 3. The section is underreamed to increase the hole diameter. The next step is to find the length of the section and diameter of the underreammed hole based upon the specifications of the deviating tool. A special packer, similar to a metal plug, is placed in the casing below the milled section. The whipstock deviating tool is connected to the drillpipe and the assembly is run into the hole and positioned at the kickoff point.

The same general orientation procedure is used as for regular deviation operations. First, turn and direction are found, and then the whipstock is oriented with these corrections and set on the packer. The slender tube is lowered with the retaining cable. The mud pump is started and circulation begins down the drillpipe, through the slender tube, and out the jet nozzle. Hydraulic mud pressure against the pressure seal on top of the tube forces it downward. The tube passes through the curved guides on the whipstock. These turn the tube through a 90° angle from vertical to horizontal. A stream of high pressure mud from the jet nozzle erodes the formation and drills the hole horizontally. When the horizontal section is drilled in this manner, the tube is pulled back into the pipe with the retaining cable. Additional horizontal holes may be drilled from the same wellbore by turning the whipstock in another direction and repeating the procedure. In most cases completion is done in the open hole without casing, using mild stimulation if required.

REGULAR SHORT-TURN

Regular short-turn horizontal patterns have a turn radius of about 30–60 ft for drilling from cased holes. There are various tools and different tool sizes (diameters), all somewhat small. Generally, longer horizontal sections are drilled with larger diameter tools, commonly several hundred feet and extended deeper in a few cases.

The process begins by sidetracking, building angle, and drilling the curved section with a special angle-building assembly. It has a nonrotating flexible tubular steel shell made of short lengths of pipe. The lengths connect together with articulated connections, similar to a hollow balljoint, for flexibility. This flexible shell carries the vertical thrust to the bit and acts as a spring to facilitate building angle. A flexible liner inside the shell contains pressure for circulating drilling fluid. An internal drive shaft, supported by bearing packs, carries torque from the drillstring to the bit. The horizontal section is drilled with a similar but longer flexible shell, without spring action and stabilized to control direction.

The hole is prepared by first removing a section of casing and underreaming similarly to the drainhole system. The deviation assembly sidetracks from a modified whipstock. The whipstock may be seated in several ways. It may be placed on a packer set in the lower casing or set inside a hole drilled into a cement plug. In either case, the whipstock is run, oriented to the correct direction with standard orienting procedures, and set. The whipstock is released, and the drillpipe is pulled out of the hole. The angle-building assembly is connected to the drillpipe and run into the hole.

Sidetracking begins by rotating the angle-building assembly as it guides off the face of the whipstock. Sidetracking and drilling continue in the curved hole section at an increasing upward angle in the direction of the whipstock face until it is horizontal. Then the angle-building assembly is pulled out of the hole. The stabilized drilling assembly is run, and the straight, horizontal section is drilled. Another version of the system rotates the bit with an articulated motor, which improves hole guidance and general deviating and drilling operations. The well is completed as an open hole or a special, flexible-type slotted liner is run.

MEDIUM-TURN

Medium-turn horizontal hole classifications commonly are drilled in the open hole. But, sidetracking in cased holes is moderately common with lower turn radius patterns of about 300 ft. The drilling program contains the well pattern, including build angles and lengths of the curved and horizontal sections (see Fig. 5–3 and Table 5–1). The pattern is drilled in the same general manner as for deviated holes with allowances made for the higher angle and angle-build rate, and using different motor assemblies in most cases. The main differences are in the selection of the deviation assembly and because of the higher drag and torque (with their associated higher risk). The type of drilling assembly depends upon the well pattern, formations, and specific hole conditions. Motor assemblies are most commonly used because of this; tangents are used sometimes. It is common to use a measurement-while-drilling (MWD) measuring system. Steering tools may be used in the upper part of the curved section but are less common. A magnetic single-shot is used a limited amount for special measurements such as verifying the accuracy of the other tools (see Fig. 5–4).

CURVED SECTION

The curved section of horizontal holes turns through a 90° curve from horizontal to vertical with an average turn radius of 300–800 ft. Deviating or sidetracking starts as described in Chapter 3. Angle-building continues with the deviation (or sidetracking) mo-

Figure 5–3
Horizontal well plan—schematic
(courtesy of Drilex Systems, Inc.)

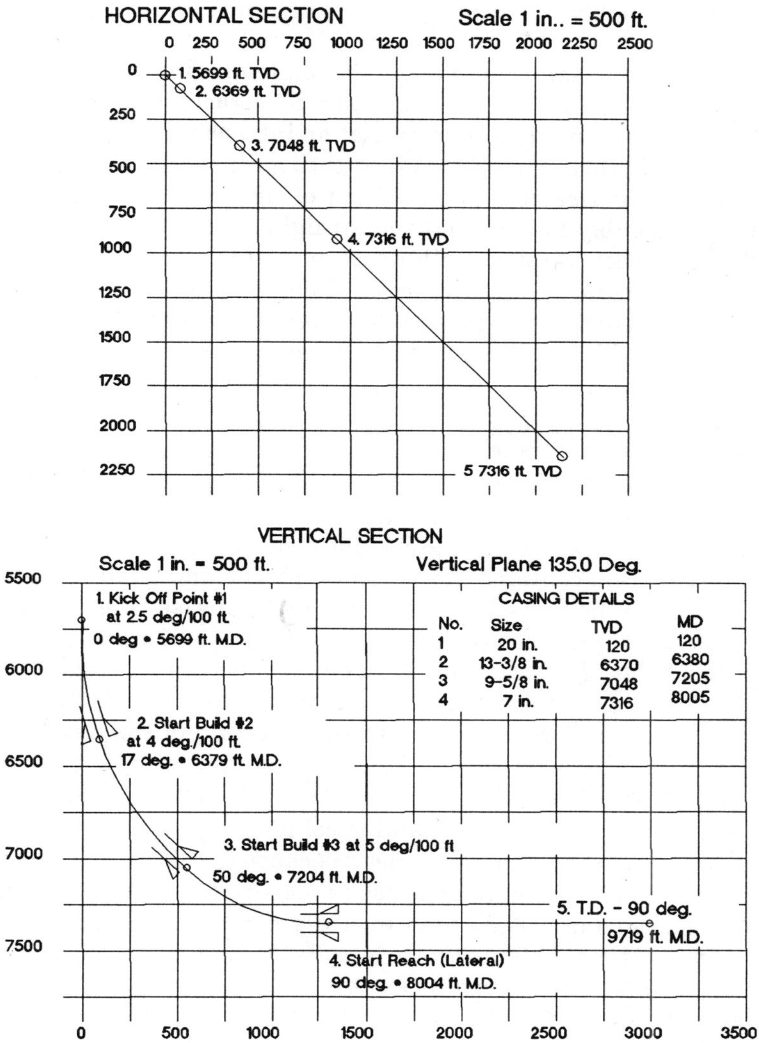

WELL PROPOSAL

PT. #	MD	TVD	VS	INC.	DIR	NORTH	SOUTH	DLS
1	5700	5700	0	0	135	0	0	2.5
2	6380	6370	100.0	17	135	-70.8	70.8	4.0
3	7205	7048	549.0	50	135	-388.4	388.4	5.0
4	8005	7316	1285.8	90	135	-909.2	909.2	0.0
5	9719	7316	2500.3	90	135	-1768.0	1768.0	0.0

Table 5–1
Horizontal Well Plan—Course Measurement Data.
(courtesy of Drilex Systems, Inc.)

Wellplan Profile

Measured Depth (ft)	Inclination (degs)	Azimuth (degs)	TVD (ft)	Departure Vertical Section (ft)	North (+) South (-) (ft)	East (+) West (-) (ft)	DLS (deg/ 100 ft)
I. Kick Off Point at 2.5°/100 ft in 17 1/2" hole.							
5699.35	0.00	0.00	5699.40	0.00	0.0	0.0	0.00
5729.35	0.75	135.00	5729.40	0.20	-0.1	0.1	2.50
5759.35	1.50	135.00	5759.40	0.80	-0.6	0.6	2.50
5789.35	2.25	135.00	5789.40	1.80	-1.2	1.2	2.50
5819.35	3.00	135.00	5819.30	3.10	-2.2	2.2	2.50
5849.35	3.75	135.00	5849.30	4.90	-3.5	3.5	2.50
5879.40	4.50	135.00	5879.30	7.10	-5.0	5.0	2.50
5909.35	5.25	135.00	5909.10	9.60	-6.8	6.9	2.50
5969.35	6.75	135.00	5968.80	15.90	-11.2	11.2	2.50
6029.35	8.25	135.00	6028.30	23.70	-16.8	16.8	2.50
6119.35	10.50	135.00	6117.10	38.40	-27.1	27.1	2.50
6179.35	12.00	135.00	6175.90	50.10	-35.4	35.4	2.50
6239.35	13.50	135.00	6234.40	63.30	-44.8	44.8	2.50
6299.35	15.00	135.00	6292.60	78.10	-55.2	55.2	2.50
6359.35	16.50	135.00	6350.30	94.40	-66.7	66.7	2.50
2. Start Build #2 at 4°/100 ft in 12 1/4" hole and 13 2/8" csg							
6379.35	17.00	135.00	6369.50	100.10	-70.8	70.8	2.50
6409.35	18.20	135.00	6398.10	109.20	-77.2	77.2	4.00
6469.35	20.60	135.00	6454.70	129.10	-91.3	91.3	4.00
6529.35	23.00	135.00	6510.40	151.40	-107.1	107.1	4.00
6559.35	24.20	135.00	6537.80	163.40	-115.6	115.6	4.00
6619.35	26.60	135.00	6592.00	189.20	-133.8	133.8	4.00
6679.35	29.00	135.00	6645.10	217.10	-153.5	153.5	4.00
6739.35	31.40	135.00	6697.00	247.30	-174.9	174.9	4.00
6799.35	33.80	135.00	6747.50	279.60	-197.7	197.7	4.00
6859.35	36.20	135.00	6796.70	314.10	-222.1	222.1	4.00
6889.35	37.40	135.00	6820.70	332.00	-234.8	234.8	4.00
6919.35	38.60	135.00	6844.30	350.50	-247.8	247.8	4.00
6879.35	41.00	135.00	6890.40	388.90	-275.0	275.0	4.00

Wellplan Profile (continued)

Measured Depth (ft)	Inclination (degs)	Azimuth (degs)	TVD (ft)	Departure Vertical Section (ft)	North (+) South (-) (ft)	East (+) West (-) (ft)	DLS (deg/ 100 ft)
7039.35	43.40	135.00	6934.90	429.20	-303.5	303.5	4.00
7099.35	45.80	135.00	6977.60	471.30	-333.3	333.3	4.00
7189.35	49.40	135.00	7038.30	537.80	-380.3	380.3	4.00
3. Start Build #3 at 5°/100 ft in 8.5" hole and 9 5/8" casing							
7204.35	50.00	135.00	7048.00	549.20	-388.4	388.4	4.00
7234.35	51.50	135.00	7066.90	572.50	-404.8	404.8	5.00
7294.35	54.50	135.00	7103.00	620.40	-438.7	438.7	5.00
7354.40	57.50	135.00	7136.60	670.10	-473.9	473.9	5.00
7384.40	59.00	135.00	7152.40	695.70	-491.9	491.9	5.00
7444.40	62.00	135.00	7181.90	747.90	-528.8	528.8	5.00
7504.40	65.00	135.00	7208.70	801.60	-566.8	566.8	5.00
7564.40	68.00	135.00	7232.60	856.60	-605.7	605.7	5.00
7594.40	69.50	135.00	7243.50	884.50	-625.5	625.5	5.00
7654.40	72.50	135.00	7263.00	941.30	-665.6	665.6	5.00
7684.40	74.00	135.00	7271.70	970.00	-685.9	685.9	5.00
7744.40	77.00	135.00	7286.70	1028.10	-727.0	727.0	5.00
7804.40	80.00	135.00	7298.70	1086.90	-768.5	768.5	5.00
7864.40	83.00	135.00	7307.50	1146.20	-810.5	810.5	5.00
7924.40	86.00	135.00	7313.30	1205.90	-852.7	852.7	5.00
7984.40	89.00	135.00	7315.90	1265.80	-895.1	895.1	5.00
4. Start Reach in 6" hole and set 7" liner & maintain direction.							
8004.40	90.00	135.00	7316.10	1285.80	-909.2	909.2	5.00
8504.40	90.00	135.00	7316.10	1785.80	-1262.8	1262.8	0.00
9504.40	90.00	135.00	7316.10	2785.80	-1969.9	1969.9	0.00
5. Hole T.D.							
9718.55	90.00	135.00	7316.10	3000.00	-2121.3	2121.3	0.00

Figure 5–4
Drilling a medium-turn horizontal well

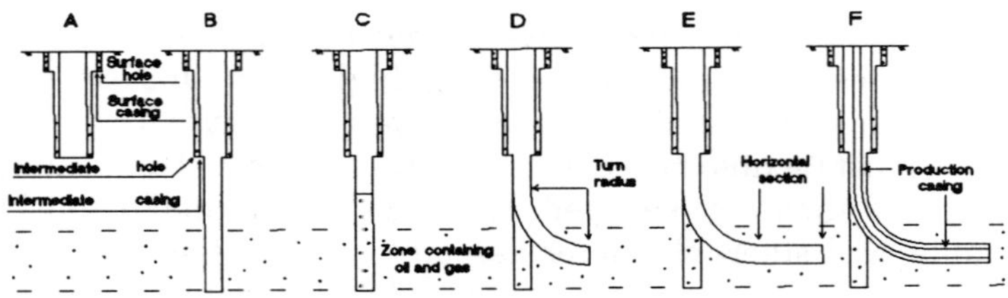

A = Vertical hole drilled and cased to surface or intermediate casing point
B - Vertical hole drilled through formation containing oil and gas (optional).
C = Vertical hole plugged-back to kickoff point.
D - Curved hole section drilled through 90° turn radius. May take two or more steps including one or more strings of casing for longer turn radius.
E = Horizontal hole section drilled
F - Curved and horizontal hole sections cased and completed. May take several steps.

tor assembly if it builds at a sufficient rate. Otherwise, the assembly is pulled and one is run that builds angle satisfactorily. Drilling inside cased holes (usually 7 in. diameter or larger casing) is done with small diameter, slim-hole tools. A section of casing is removed by milling, a sidetracking plug is set and dressed off, and sidetracking continues as described in Chapter 3. Sidetracking off a whipstock is not recommended. Some operators prefer to sidetrack out of cased holes with a lower build rate, increasing the build rate after drilling part of the curved section. This reduces the risk of keyseating near the bottom of the casing.

Drilling and angle-building operations continue, measuring drift and direction periodically. Trips are made as necessary to change the assembly, replace a dull bit, and for similar reasons. A tangent section may be drilled if required. The drillstring is pulled and a steerable or sometimes hold assembly is run and oriented. A straight, inclined hole section (tangent) is drilled to the required depth. Then the drillstring is pulled, a directional motor assembly is run, and angle-building continues in a smooth, upward curve until the hole is horizontal.

The curved section of the hole may be cased before drilling the horizontal lateral, or both may be cased together. Running of the casing depends upon the turn radius, length of the horizontal section, formation conditions, torque, and drag. Inadequate bit weight normally is not a problem in the curved section except near the end at higher angles. If it is, the bottomhole assembly may be split as described in the following section about drilling the horizontal section of medium turn radius holes.

Drilling is done with motor assemblies because rotary assemblies usually cannot build angle at these high rates. One exception is the hooligan rotary assembly, used in a few holes with longer turn radii. The most flexible motor assembly that will build angle at the required rate should be selected. These include normal deviation, adjustable, and steerable motor assemblies. The steerable motor assembly has many advantages as described in Chapter 2. A fixed, three-point support assembly builds angle at the highest rate. The angle-build rate is predictable and effective at higher angles when angle build and directional control is a problem.

HORIZONTAL SECTION

Horizontal sections of medium-turn holes have angles of about 90° and vary from 75°–100° depending upon formation conditions and well patterns. The horizontal section is drilled with either a hold or low-angle-build steerable motor assembly. The hold assembly may have a limited amount of stabilization. Often a nearbit stabilizer gives a slight angle build and counteracts the angle-dropping tendency.

The steerable assembly is drilled with often, because angle changes generally are small and within the angle control capability of the assembly. There are also the advantages of drilling straight ahead or controlling the direction of the horizontal hole without making a trip to run a corrective assembly. A common steerable assembly has a bent housing with a low angle-bend of 0.25°–0.5°, possibly with a very thin deflection pad, to prevent motor housing wear.

The drillstring is pulled out of the hole after drilling the curved section. A motor assembly is run and drilling of the horizontal section begins. Reaming normally is unnecessary because the assemblies are relatively limber. Reaming can be done with a nonaggressive reaming assembly, if required. Reaming should be done very carefully because of the high risk of unintentionally sidetracking in the highly deviated hole. Drift and azimuth measurements should be recorded periodically. Trips should be made as necessary for such reasons as to change bits and to change or modify the assembly. Drilling continues, and the horizontal section is completed.

Applying sufficient bit weight for an optimal penetration rate is often a problem, especially at higher angles and while drilling the horizontal section. Conventional bit weight for efficient drilling is about 2,000–5,000 lbs per inch of bit diameter. Available bit weight from a given assembly theoretically is reduced by a factor related to the cosine of the drift angle. The cosine approaches zero as the angle of the hole approaches 90°. Motor assemblies drill efficiently with less bit weight than rotary assemblies. They compensate for reduced bit weight with the higher rotational speed of turbines and motors. Bit weight may be increased by reducing drag and torque conventionally as described earlier in the section on drag and torque. But, often this is not sufficient for an optimal drilling rate in holes with higher angles and in horizontal sections.

Bit weight is often increased by dividing the bottomhole assembly into two parts. The drillstring is pulled and a split assembly is run (see Fig. 5–5). The lower part of the assembly, including the bit, motor, directional control tools, and the nonmagnetic collars are left at the bottom of the drillstring. The remainder of the drill collars are placed in the vertical hole or in an upper-curved hole that has a low drift. The two sections are connected with compression pipe or sometimes heavyweight or regular drillpipe. The drillstring is completed in the conventional manner with drillpipe from the top of the upper section to the surface. Split bottomhole assemblies reduce drag and torque so that more weight can be applied to the bit for drilling faster.

The heavier drill collars are more effective placed in the vertical hole section as compared to placing them in the horizontal or highly deviated hole section. They exert more downward force to the lower drillstring and less force on the side of the wellbore. The force is transmitted by the compression or drillpipe to the lower half of the assembly in the high-angle or horizontal hole section. Part of the downward force is still lost due to some drag and torque. The compression pipe or drillpipe connecting the two assembly sections together operates in compression so that the risk of failure increases. This apparently is an acceptable risk based on operational results. Also, there apparently is less risk of failure while drilling with the drillstring stationary and rotating the bit with a motor or turbine compared to conventional drilling by rotating the entire drillstring. Drilling with a steerable assembly and rotating the drillstring slowly also is acceptable. It is important to operate the drillstring carefully in all cases.

The jar-bumper sub is either run on the lower half the assembly or omitted, depending upon hole conditions. The jar bumper requires three to five drill collars immediately above it for effective

Figure 5–5
Split assembly

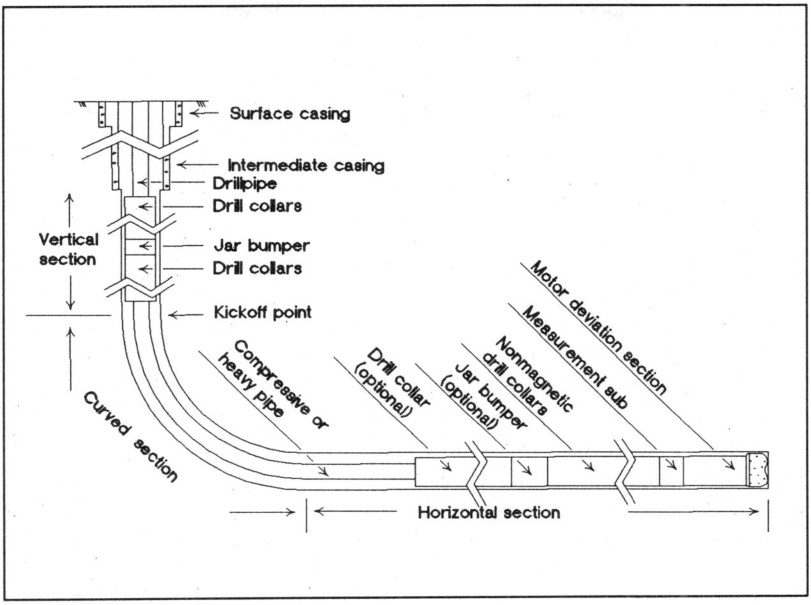

jarring and bumping. This increases the weight in the lower assembly. One alternative is positioning the jar-bumper sub on top of the lower half the assembly and let the compression pipe provide weight for the jarring action. A jar-bumper sub should always be placed near the top of the upper half the assembly.

LONG-TURN

Long-turn horizontal hole classifications are drilled mainly by deviating in open holes. Wells in this classification are characterized by larger hole sizes and are very susceptible to high drag and torque because of long open hole sections. Holes sizes range up to 12 1/4 in. diameter, although smaller diameter holes are more common. Larger tools sizes restrict the pattern to new wells and a few old wells with large diameter casing. There is a wide variance of turn radii within this classification, since it includes most holes between very high directional and medium-turn patterns. This is a common method of horizontal drilling.

Deviation assemblies for long-turn patterns may be more flexible than the medium-turn patterns. Actual assembly selection is dependent upon the well pattern, formations, and specific conditions in the well. Larger positive displacement motors with higher horsepower ratings are used. Turbines are more common, especially in offshore operations where longer turn radii are common. Normally the measurement-while-drilling (MWD) measuring system is used. Some operators use steering tools in the upper part of the curved section; the magnetic single-shot is used in a few cases, but MWD is the most common. Longer curved sections often require various casing strings.

CURVED SECTION

The curved section of horizontal long-turn holes turns through a 90° angle from horizontal to vertical with an average turn radius of 1,000–3,000 ft. The first step is to deviate and begin drilling the curved section with a standard deviation assembly as described in Chapter 3. The same general procedures are used for drilling the high-angle directional and medium-turn patterns. Reduced angles of build and longer open hole sections must be allowed for. It is necessary to establish curvature and then drill with one of several assembly options (see Fig. 5–6).

Curved sections of shorter-turn patterns are drilled similarly to drilling medium-turn patterns with a longer turn radius. Drilling continues with the deviation assembly if it builds angle within the specifications of the well plan. Otherwise, an assembly is selected to build angle according to the well pattern, including overcoming the influence of the formations on building angle and controlling hole direction. Most commonly this will be a multipoint contact or steerable motor assembly. Drift and direction are measured periodically to ensure that the well plan is followed and trips are made

Figure 5–6
Drilling a long-turn well

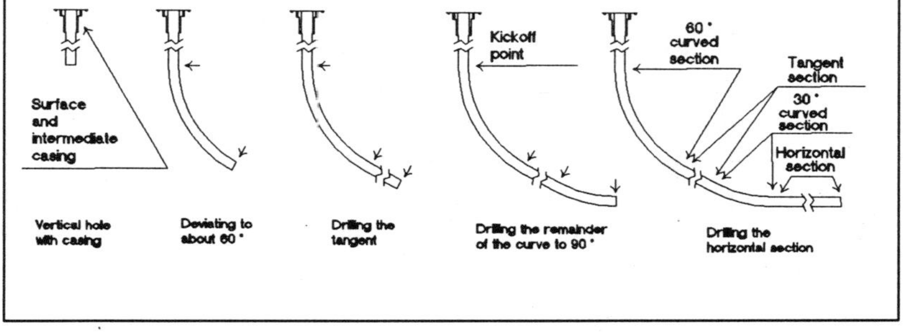

as needed. Drilling continues in this manner until the hole is horizontal.

Patterns with longer turn radii are drilled using directional drilling techniques in the earlier part of the curved section. This is more common where bit walk and angle building is less of a problem. Drilling continues with the deviation assembly until the angle is about 20°. Then an angle-build rotary assembly is run and drilling continues, building angle to about 60°. Drilling continues to higher angles with rotary assemblies in a few cases, such as a hole with a very long turn radius. Normally, rotary assemblies are less efficient for building angle and controlling direction at higher drift angles. An aggressive hooligan assembly is a good angle-building assembly under favorable conditions. The next step is to ream, correct hole drift and direction, and make trips for modifying or changing assemblies and to replace worn bits.

The remaining curved section is normally drilled from 60° to horizontal with motor assemblies, drilling with steerable motor assemblies as often as possible. The deviation equipment is arranged on the steerable assembly for building angle a few degrees higher than the pattern design build rate. It also is possible to drill alternately with the motor, but only for controlling drift and direction and while rotating the drillstring for drilling straight ahead. The planned curve should be maintained in this manner, possibly eliminating tangent sections. Steerable assemblies may not build angle at a satisfactorily high rate, so a regular angle-build motor assembly is run.

Various motor assemblies build angle reliably at high angles. These include those with double contacts, such as the bent sub and bent housing, fitted with deflection pads if needed, or a double universal motor housing with stabilizer guides. Drilling continues in this manner until the hole reaches the horizontal.

Tangent sections should be drilled as needed. The procedure generally is similar to drilling a straight, inclined section in a directional hole. Tangents are often placed at an inclination of 60°. Tangents are omitted in some holes that have longer turn radii, because drilling longer sections provides time for more well path adjustments. The steering tool makes small, correcting adjustments under favorable conditions.

Intermediate-turn curvatures are drilled with either one of the two methods described, or both. Drag and torque increase substantially at higher angles and in deeper holes; in severe cases, this decreases available bit weight. It is not uncommon to set casing because of this. Motor assemblies aid in reducing torque, since the drillstring does not rotate. This may help increase available bit weight.

HORIZONTAL SECTION

Horizontal sections of long-turn holes have angles of about 90°, varying from 75°–100° depending upon formation conditions and well patterns. The horizontal section is drilled with either a hold or low-angle-build steerable motor assembly. Procedures similar to drilling the horizontal section of the medium-turn pattern are used. Rotary assemblies are seldom used. Drag and torque increase with increasing depth. Torque may approach the maximum limiting torque strength of the drillstring in very deep (measured depth) holes. This has occurred even in a completely cased hole. Motor assemblies should be used here, since they do not require rotating the drillstring. Drag and torque may be minimized with the correct type, high quality mud system and other actions as previously described.

Split assemblies are generally omitted in long-turn patterns because of the long distance between the upper and lower parts of the assembly. For example, a turn radius of 2,000 ft means a separation of 1,350 ft of compression pipe between the upper and lower parts of the assembly. This is when the bit starts drilling the horizontal hole with the upper part of the assembly in the vertical hole. The safe length between the two sections of the assembly is determined based on equipment and hole conditions. In this case, drilling should probably continue with the regular assembly until the bit weight is too low. Then the split assembly should be tried. Another alternative is to reduce the weight of the bottomhole assembly by eliminating some drill collars and using compression pipe, which reduces the length of the upper part of the assembly. Various procedures are used and drilling is completed in the horizontal section.

Barite sag occurs under favorable conditions when barite particles dispersed in the mud system begin settling or collecting and stacking on the low side of the inclined well bore. Most settling occurs when the system is without circulation, such as during tripping, while recording measurements, and during other wireline operations. The barite particles then slump or slide down the wall of the hole and accumulate similarly to cuttings. The accumulations disperse into the mud when circulation resumes, causing areas of high- and low-density mud.

This phenomenon often is not recognized or known. It may be detected by weighing the mud at short intervals while circulating bottoms-up after a trip. Mud weight variations greater than 4 PPG have been observed. These variations can cause stuck tools and lost circulation, and increase the risk of a blowout. Problem severity increases with time, higher angles of inclination, lower quality muds, and higher weight mud systems. Preventive action should be taken based on the potential cause. High quality mud with good gel properties helps retain barite particles in suspension. Noncirculating time should be minimized. Another aid is to stop periodically and circulate while tripping into the hole if necessary. Shorter intervals and increased circulation times may be necessary for more severe cases. Similar precautions should be observed while running casing.

Barite sag is not significant in vertical and low-angle directional holes, but it can cause problems in high-angle and horizontal holes, especially when using heavyweight mud systems.

EXTENDED-REACH AND COMBINATION PATTERNS

Extended-reach patterns generally are deep (measured depth), with long horizontal displacements between the surface location and the bottom of the hole. Patterns are similar to directional build-and-run wells, but they have longer and higher angle straight, inclined sections. Angle-build rates range from the higher rates found in directional patterns to those used in long-turn horizontal wells. Similarly, straight, inclined sections have a wide range of drift angles. Curved section lengths vary depending upon the well pattern. The two extremes are a medium-turn curvature with a long straight, inclined section and a very long-turn curvature pattern with a short, straight, inclined section. Intermediate variations are combinations of these. The long-turn curvature also effectively serves to provide extended reach, but it is a different basic pattern. They are common, especially offshore.

Extended-reach hole sizes range upward to 12 1/4 in. diameter although smaller diameter holes are more common. Casing programs may require larger hole diameters, usually in the shallower sections. A 12 1/4 in. hole is drilled and then opened to the required size with a hole opener in most cases when this happens. Generally, the larger holes sizes restrict the pattern to new wells and few old wells with large diameter casing. The hole is started by deviating from a vertical, uncased wellbore. The next step is to build angle and drill a smooth, upward curve with assemblies and procedures used for either deeper directional drilling or long-turn patterns, whichever has the most similar angle-build rate.

The straight, inclined section is drilled with a hold or steerable assembly. Motor assemblies are mostly used, but hold rotary assemblies may be more efficient at lower drift angles for deeper wells. General conditions and problems are similar to those encountered while drilling the horizontal section of long-turn holes. Drilling may be more difficult for very long, straight, inclined sections. Excess drag and torque can be severe and sometimes may limit maximum depth. Split assemblies are less common.

Combination patterns are mixtures of various sections of the patterns listed. Extended-reach may be combined with horizontal patterns. A medium or long-turn angle-build section is drilled after the straight, inclined section, then drilling continues horizontally. Other combinations have similar variations. Complex patterns are less common and include multiple bends and turns. They can create difficult, high-risk drilling situations.

FORMATION EVALUATION

Formation evaluation is the process of examining the formations to determine if they contain oil and gas and often provides a means of estimating production rates and reserves. This is an important procedure, since the reason for drilling the well is to find oil or gas in commercial quantities. Some horizontal well formation evaluation procedures are the same as in most other wells. Others are similar but modified for high-angle and horizontal holes. Most of these procedures record the same type of data but by different methods because of the high-drift angles.

Regular well or mud logging is common and used on most horizontal wells. The procedures are basically the same as for other wells. Lag times are longer and may be more difficult to find because of extended circulation times. Drill cuttings generally are smaller in size because of the deviated hole and for other reasons as described in Chapter 4. Analyzing small cuttings is more difficult. This requires a careful analysis and more expertise by the mud loggers but otherwise is not a severe problem.

Equipment is available for open hole testing, but is seldom used because of the high risk. The information is available by other methods, such as wireline formation tests. Open-hole tests are seldom used in high-angle and horizontal holes because of the risks and problems. Similarly, equipment is available for conventional coring. High-angle and horizontal wells are cored with a stabilized core barrel motor assembly. A shorter-length core barrel is used for holes with a shorter turn radius. Coring should be limited, because it causes loss of angle, requires extra tripping, and increases the risk of sticking. Wireline cores often replace regular cores subject to problems of running the tools into the high-angle and horizontal holes.

Electric well logging is the main method for evaluating formations during and after drilling. Electric is a slight misnomer, since it refers to the method of transferring logging data to the surface by electrical means through insulated single-strand and multistrand shielded electric logging cables (cables for short). Conventional data include electrical resistivity, spontaneous potential, sonic, gamma, and neutron radiation. Standard logging instruments are run into vertical and directional holes for conventional electric logging. However, tools run on cables will not fall freely through holes with drift angles greater than 60°. Tool and cable drag overcome the force due to gravity at these higher angles. Therefore, high-angle and horizontal holes require special equipment and procedures for well logging and other wireline operations (see Fig. 5–7).

All of the available well-logging methods have advantages and disadvantages. The disadvantages tend to outweigh the advantages, especially in long, high-angle, extended-reach and horizontal sections. Measurement-while-drilling has gained wider acceptance as a well logging evaluation tool because of this.

Figure 5–7
Logging horizontal holes

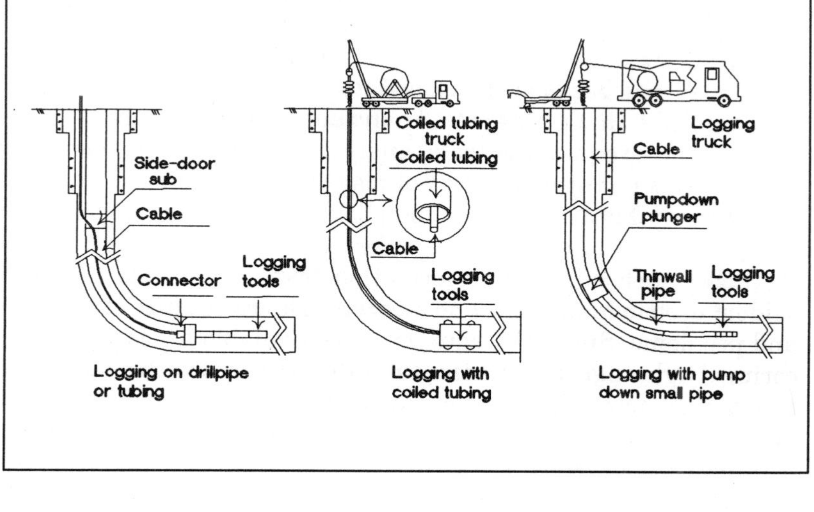

LOGGING ON DRILLPIPE

Logging on drillpipe is a procedure for logging high-angle and horizontal holes. The procedure operates somewhat similarly to the parallel steering tool measurement system. Logging tools are connected to the bottom of the drillpipe and run partway into the hole to allow for connection of a side-entry sub later. The depth should be selected so that it will be in the cased hole and above high-angle hole sections when the drillpipe and logging tools are at the bottom of the hole. This protects the logging cable from damage. The end of the logging cable has a sealed connector for connecting the cable to the top of the logging tools. The logging cable is on a reel on a logging truck that also contains the surface logging instruments. The cable is placed in the drillpipe and lowered until the cable connects with a matching receptacle on top of the logging tools. The cable may be run with weight bars or pumped down with drilling fluid. In the latter case a sealing assembly is located near the connector and a pack-off seal is used at the surface (see Fig. 5–7). The specific method depends upon the equipment in use and wellbore conditions.

A side-entry or ported sub is connected in the drillpipe string. The logging cable is passed through the ported sub so that it is inside the drillpipe below the sub and outside the drillpipe above the sub. Running of the drillpipe into the hole is resumed until the logging tools are at the bottom of the hole. The logging cable is lowered down the annular space outside the drillpipe simultaneously. Then the drillpipe and logging cable are pulled slowly while logging (recording data) with the logging tools. After logging, the drillpipe and logging tools are pulled out of the hole by the reverse procedure. Logging is accomplished through short-turn sections with articulated logging tools. Available pumpdown systems may use a different type of equipment but generally are similar in operation to that described.

This method of logging is slow and somewhat tedious. There is some risk of sticking the drillstring depending upon hole conditions. Still, the procedure permits logging the entire curved and horizontal hole sections.

LOGGING ON COILED TUBING

Logging on coiled tubing is a procedure for logging high-angle and horizontal holes (see Fig. 5–7). Coiled tubing, spooled on a reel, contains a regular insulated, multistrand logging cable inside the tubing. The logging tools are connected both to the cable and the end of the coiled tubing with a special connector. The coiled tubing carrying the logging tools is lowered into the hole. The coiled tubing pushes the logging tools into the horizontal hole. Articulated logging tools are used for curves with a short turn radius. The top of the cable connects to a device on the hub of the reel for connection to a logging truck. The hole is logged while going in the hole or pulling out, or both, depending upon the tools.

The distance that can be logged is limited, especially with heavy logging tools. The limber coiled tubing can only push the logging tools so far before buckling, causing a fishing situation. The length of horizontal hole that can be logged depends upon various factors. These include the strength and size of the coiled tubing, weight of logging tools and cable, radius of curvature, and drift angle. Chapter 2 contains the description and operation of coiled tubing.

PUMPDOWN LOGGING

Pumpdown logging is a procedure for logging in high-angle and horizontal holes (see Fig. 5–7). The logging tools are connected to the bottom of a long section of several hundred feet of small diameter pipe. The logging cable extends up through the pipe and to the cable reel on a logging truck. The upper end of the pipe has a pressure seal for sealing around the logging cable and between the small pipe and the inside of the drillpipe or tubing.

In the logging operation, the first step is to lower open-ended drillpipe into the hole so that the bottom is at the top of the section to be logged. It's normal to position the drillpipe so that the lower end of the small pipe touches the bottom of the hole when it is extended. A manifold is connected to the top of the drillpipe; the small pipe and logging tools are placed inside the drillpipe and lowered with the logging cable until they stop. A surface pressure pack-off is installed to seal around the logging cable. The drillpipe is pumped down slowly, pushing the small pipe with the logging tools attached to the bottom. Pumping continues until the top of the small pipe is at the bottom of the drillpipe. The small pipe and logging tools extend into the open hole. Logging data are recorded while pulling the small diameter pipe and logging tools with the logging cable back into the drillpipe.

Another section of the hole is logged by first pulling the logging tools and small pipe out of the drillpipe. Then the drillpipe is pulled out of the hole a distance slightly less than the extended length of the small pipe. The small pipe and logging tools are lowered and the procedure is repeated to log the section of hole. It is common to overlap the sections logged for continuity. The length of small diameter pipe that can safely be run limits this method of logging.

The procedure is slow and tedious to operate, and the small diameter pipe is subject to breaking.

ROLLER LOGGING

Roller logging tools were an early development for recording logging measurements conventionally. The tool containing logging and recording instruments has multiple sets of three rollers, usually with a diameter of 6 in. They are positioned about 120° apart, spaced evenly along the length of the tool. As the tool is lowered with the logging cable, it rolls down through the inclined casing. The procedure is effective at angles up to about 65° or slightly more. This is subject to depth and the length of logging cable dragging behind the roller logging tool. The procedure was used extensively for logging early extended-reach wells. A later improvement included a seal assembly on top of the logging tools for pumping them down the hole through special tubulars.

CASING AND CEMENTING

All horizontal holes should be cased except for some with very short-turn radii and those designed for an open-hole completion. Running liners is a common practice, especially in horizontal sections. The holes should be cased for the same reasons as in vertical and directional holes and for special conditions required in horizontal drilling. Casing programs should be included with the well plan. Standard casing design should be used with allowances for casing wear, hole curvature, drag, torque, and other hole problems. It is important to consider completion and stimulation when designing the casing program.

Various factors determine the length and number of casing or liner strings, depending upon the well pattern and specific conditions in the well. Openhole sections should be limited to a maximum length of about 12,000 ft in vertical holes under optimum conditions. A much shorter openhole section should be used in curved and horizontal hole sections where drag, torque, and other hole problems are common. Casing should be set to reduce problems caused by high drag and torque and to protect and save the existing hole when drilling problems occur. This also may reduce problems while drilling deeper that are caused by problem formations in upper sections. It may be necessary to further reduce the length of open hole if formation or hole conditions are unfavorable.

Problem formation conditions include fluid sensitivity, spalling, fracturing, geopressuring, and lost circulation. Hole condition problems include crooked hole, keyseating, wall sticking, and high drag and torque. Casing should be set near the top of the pressure transition zones where the formation pressures change rapidly over a short depth interval. It is common to set casing through shallower productive zones to protect and save them while drilling to deeper productive zones. Normally, casings are set through producing zones unless completing in the open hole.

Casing programs for curved and horizontal sections depend upon the casing program above the kickoff point. This is a variable that affects casing setting depths and depends upon specific conditions in wellbore, the depth of the kickoff point, and the subsequent deviation program. Casing above the kickoff point is important but cannot be included here because of unknown factors (see Fig. 5–8).

MEDIUM-TURN

Medium-turn holes with casing set near the kickoff point require one to three casing strings below this point. The following casing programs are possible:

1. *Drill the curved and horizontal sections and set casing at the end of the horizontal section.*
2. *Drill the curved section and set casing. Then drill the horizontal section and set casing.*
3. *Drill the first half of the curved section and set casing. Complete drilling the curved section and set casing. Drill the horizontal section and set casing.*

Figure 5–8
Casing designs

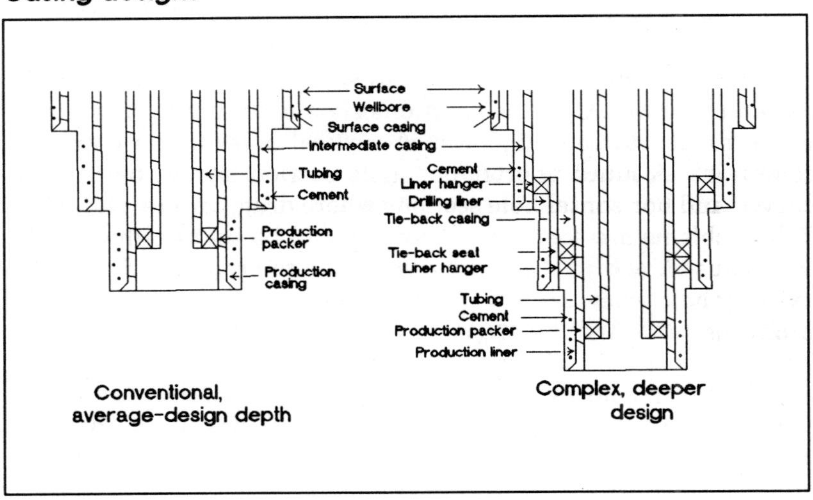

The third casing program is common for wells with casing set some distance above the kickoff point. It also helps to reduce drag and torque problems while drilling deeper.

It is common to run liners instead of casing in the horizontal section of medium-turn holes. Perforated, slotted, or wire-wrapped (screen) liners are used for open-hole completions. Some horizontal sections are completed without casing or liners. This may be acceptable if the formations are hard, competent, and completely stable, but even then the procedure is questionable. Some holes apparently have closed after a period of time, even in hard formations and probably due to caving formation material. It is best to always run casing or a liner in the open hole if there is any possibility of the hole closing later.

Sometimes the upper casing may be severely worn while drilling deeper. A complete string of casing should be run to cover these instead of a lower liner, or a tie-back casing may be used with a lower liner. This type of casing program is applicable to high-pressure wells.

LONG-TURN

Casing programs for long-turn holes are more varied because the holes are deeper. The same general considerations apply as for medium-turn radius holes. Wells with a shorter turn radius may have casing programs similar to medium-turn wells. Wells with longer turn radii have programs similar to vertical and directional wells. The final casing program depends upon the conditions in the specific well.

An average long-turn well with casing some distance above the kickoff point may have casing through about one-half to two-thirds of the curved hole. Then drilling of the curved and horizontal sections is finished and casing is run again. Holes with longer turn radii and some deeper holes may have casing set at an angle of about 60° and at the end of the curved and horizontal sections. Tie-back casing and stub liners are run in special cases. A liner should be set through the horizontal section. Then tie-back casing is run from the liner top through the curved section and to the surface. This isolates all older and possibly worn casing, especially in the curved sections. It also is a common procedure for wells stimulated with high pressures at completion or wells in high-pressure formations.

A less common alternative places stub casing from the liner top through the curved section. Generally, a hydraulically set liner hanger is used. Setting mechanically set hangers requires rotating the drillstring. This may be difficult or nearly impossible in high-angle or horizontal holes. It will be necessary to provide for an extra casing for higher risk holes and where formation and hole problems may be more severe than projected. Casing programs for extended-reach holes often are similar to long-turn programs.

CEMENTING

Cementing procedures in high-angle and horizontal wells are similar to cementing vertical and directional wells with added precautions due to the directional and horizontal hole. Two of the main problems are designing the slurry and displacing it correctly. Most cement service companies have computer simulation programs to help design the cement program, including displacement. It is important to observe good cementing practices.

The conventional method of isolating formations is to cement them correctly and prevent channeling. This helps to confine the formation fluids, preventing them from migrating into producing zones from adjacent formations. It ensures better testing and stimulation results. Isolation also helps control formations with different pressures, water encroachment, and different types of production like oil and gas.

Isolation may be accomplished with good bonding between the casing and cement and between the cement and formation. Sometimes this is difficult, especially in horizontal hole sections, but it is necessary for successful completions. Bonding may be checked with a cement-bond log, depending upon the specific situation. Correcting it later by conventional methods of perforating and squeezing is difficult and risks increase.

It is important to use good quality cement. A correctly designed cement slurry allows for good flow properties, correct thickening time, and ultimate compressive strength. The cement can be tested in the standard manner. Additional testing often is necessary. The slurry should be tested for sedimentation or settling. Some sediments in the slurry tend to settle a small amount, and this normally is not a problem in vertical holes. It can be a problem in highly deviated or horizontal holes because of the relatively short distance for particle movement before it contacts the wall of the hole. The net result is a thicker, denser slurry on the low side and a thinner, lighter slurry on the high side. The slurry should be tested for free water, which can accumulate on the upper side of the hole after the slurry is in place. This creates channels outside the casing, preventing good isolation. Water-free cements are available and give good results.

It is normal to batch mix the cement slurry. Spacers should be used as needed to ensure compatability with the slurry and mud. Settling of weighted spacers should be checked for if this is a possible problem. Spacers should be designed to water-wet the formation if oil mud was used. Spacer volumes and flow rates should be calculated so that the spacers contact the formation for the recommended period of time. Good slurry flow properties and correct displacement procedures prevent the trapping of mud and help to displace it completely with the slurry.

The importance of a good mud system cannot be overemphasized. It is vital to maintain it in good condition with the recommended yield, gel strength, fluid loss, and otherwise overall good rheology. Yields may be increased for better hole cleaning in higher angle and horizontal holes. Good mud cleans the hole better and is more easily displaced with the slurry. Trapped mud (pockets of undisplaced mud) causes subsequent channels. The hole should be circulated as necessary to ensure that it is clean.

Solids, normally located on the low side of the hole, are difficult to remove with the slurry and remain as an area of weak or ineffective bonding. Cement should be displaced at turbulent flow rates when possible. It is important to try to maintain a constant flow during displacement, without stops. This maintains shear rates, so that the mud does not rebuild gels. Circulation should be kept continuous while dropping plugs by automatic plug-launching equipment.

Correct mud removal and cement slurry displacement is a major concern. Besides slurry flow and related properties, there are other items that affect the success of the cement job. These include installing the correct equipment on casing or liners and movement during cementing. Casing and liners lie on the sides of directional and horizontal holes. Centralizers provide standoff, holding the casing away from the wall of the hole. This allows a continuous sheath of cement around the centralized casing for improved cementing.

Various types of centralizers include solid body, rigid, and bow spring. Each has advantages and disadvantages; the selection should be based on common usage in the area and the specific conditions in the well. The number of centralizers needed is found by calculating standoff with computer programs (see Fig. 5–9). Standoff is the distance from the outer wall of the casing to the inner wall of the hole, expressed in inches or as a decimal fraction. The fractional standoff at a specific point is found by dividing the

Figure 5–9
Standoff

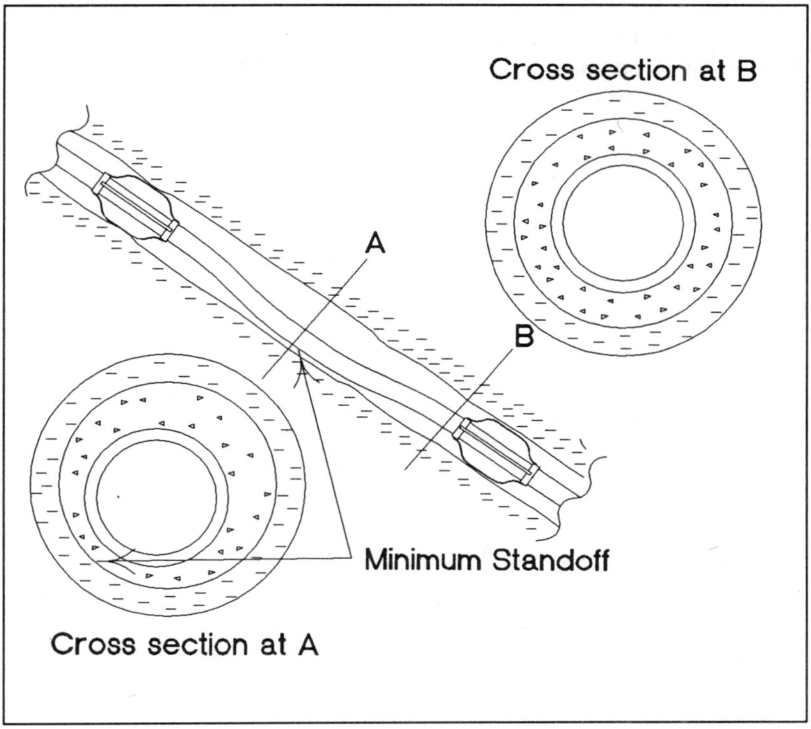

standoff at that point by the sum of the hole diameter minus the outer casing diameter. Scratchers remove wall cake and improve cement-to-formation bonding.

As a note of caution, additional equipment installed on casing and liners increases the risk of sticking. Stuck casing in horizontal holes is extremely difficult to release. Often it cannot be cemented in place efficiently because it cannot be moved and circulation may be restricted. Also, it may be so small that another string of casing cannot be run through it to case the remaining part of the horizontal hole. It is important to consider this when designing the well.

Rotating and reciprocating casing and liners can improve cementing efficiency due to better mud displacement and cement bonding. The lower section of the casing is rotated in the highly deviated holes with special equipment. Liners are rotated during cementing for similar purposes and to reduce leakage around the top of the liner. There are various opinions about rotating liners in the industry. It unquestionably improves drilling fluid displacement and improves bonding, yet it can be a high-risk operation.

Inflatable (expandable) cement-filled packers improve zone isolation. The inflatable packers are placed at points such as productive formation boundaries where good isolation is needed. After cementing the casing, the packers are inflated and filled with cement. Inflatable packers are successful under various severe operating conditions where isolation by conventional cementing is very difficult. The California Division of Oil and Gas accepts inflatable packers as a water shutoff tool and waives subsequent water shutoff tests if the packer seats properly.

Cement filtrate and spacers damage some fluid-sensitive formations. This restricts the flow of oil and gas from the formations and in severe cases almost completely blocks the flow. Special precautions can prevent formation damage. A viscous, nondamaging fluid, usually a polymer, is placed in the horizontal hole before casing and cementing the upperhole section. This covers and protects the productive formation. An internal seating device (slip-and-seal assembly receptacle) is connected to the bottom of the casing. The casing is run to the top of the horizontal section or the sensitive formation and cemented. A liner is run and the protective fluid is removed by circulating with a nondamaging fluid. Then the top of the liner is seated in the casing seating device. The well is allowed to flow. The flow may be stimulated lightly by circulating a chemically treated fluid to wash and clean the face of the formation.

An alternative procedure includes running casing to the bottom of the horizontal hole. An inflatable packer and cement diverting tool are placed above the sensitive formations. The upper section of casing is cemented making sure that the packer has sufficient strength to prevent the cement slurry from moving downward if the lower formations are weaker and take fluid.

COMPLETIONS

Generally, horizontal wells are completed similarly to vertical and directional wells, but modified as necessary for the horizontal hole. Completion operations must be conducted in order to optimize production rates.

Many conventional completions use tools run on wirelines and cables. Some of these include perforating, production logging, setting packers and plugs, and fluid-flow surveys. These often are difficult to accomplish in horizontal completions, similar to the problems of electric logging. Completion tools usually are run on wirelines by the same procedures as logging tools, as described earlier in Chapter 5. The logging tool is effectively replaced with a completion tool and run on coiled tubing or regular tubing instead of drillpipe. These types of operations should be kept to a minimum because of the time and risk involved. Completions should be modified as necessary when these procedures cannot be used.

The wellbore normally is prepared as the first completion procedure after cementing. The liner top is tested by pressure and inflow, and leaks are repaired by squeeze cementing. Tie-back casing should be run and cemented at this point, if used. Normally the tubing is run to total depth, sometimes cleaning the walls of the casing and liners with a casing scraper. The hole is then circulated until it is clean. Some wells use the regular mud as a completion fluid. Others use a special fluid; if this is the case, it should be displaced into the hole at this time. The remaining procedures depend upon the specific type of completion.

An open-hole completion is common and generally uncomplicated. Almost any type of horizontal well may be completed in this manner, but it is more common in medium-curvature horizontal wells. Still, there always is a latent risk of caving and plugging at a later date with this completion. Tubing is run to the bottom and the hole is circulated clean. The formation face is washed with a light flush as needed. Wellhead controls are installed and the well is placed on production. It is an easy, fast completion with little risk.

Fracturing and acidizing is less common. It is difficult to determine the amount and location where treating fluid enters the formation. One possible stimulation here is matrix-acidizing in carbonates. Otherwise, fracturing, acidizing, or acid-fracturing treatments are less common.

Sometimes there are divided opinions about where to position the bottom of the tubing in horizontal completions when sand production is a potential problem. One opinion is that produced sand (the small amount of sand that enters the wellbore over time) will flow out of the hole with produced fluids. So the tubing or tailpipe is run to the bottom of the hole or below the deepest perforated interval. This keeps the hole open and clean.

A divergent opinion takes the position that produced sand will ultimately collect around and stick tubing set deeper in the hole, below producing perforations. So the bottom of the tubing is positioned at the top of the formation or above the perforations. However, there is a risk in this case that the sand will settle down into the lower hole and plug part or all of the producing formation.

Generally the best procedure is to place the tubing deeper in vertical and low-angle directional wells. Placement for horizontal holes depends upon specific well conditions. It is best to see what operating results are common in the area.

Placing a liner in the horizontal section is a common and often preferred method of completion. There is less risk of hole closure due to caving and plugging. Some cemented liners are similar to regular casing and require perforating as described later in this section. Excluding this, the type of liner is selected based upon specific well conditions. A preperforated or slotted liner is selected for oil and gas production if it is suitable for stimulation requirements and sand production is not a problem. These wells are completed by running tubing, cleaning out, stimulating as needed, and then placed on production.

Many wells produce sand, including sediments and fine particles, in varying amounts. Sand production is a problem in some horizontal wells. Horizontal wells often produce greater volumes of sand because of pressure differentials and the longer section of producing formation exposed to the wellbore. Overall flow rates in the horizontal section may be less, so the fluid has a lower sand-carrying capacity. Even very small amounts of sand, especially larger-size sand grains, tend to fall to the bottom side of the wellbore. They can accumulate, plugging the hole. The problem is somewhat comparable to cleaning the hole with low circulation rates during drilling. The position of the bottom of the tubing may or may not solve the problem.

It is possible to restrict sand production and reduce related problems by placing one of several types of slotted liners or screens in the wellbore opposite the producing section (see Fig. 5–10). A preslotted liner (various size slots are available) is the most economical but least effective method of controlling sand entry. Wire-wrapped screens or prepacked liners are better methods of controlling sand production in applicable cases.

It is necessary to analyze a sample of formation sand in order to select the correct size of holes in the screen (openings). Screen openings should be sized so that only very small sand grains pass. This sand moves to the surface as fine sediment with produced fluids. Modifications of this liner are run in open or cased holes. For placement in cased holes, the special liner or screen is connected to the tubing, run into the hole, and placed opposite the perforated section. It is run similarly in open holes, usually using a packer or by seating the top into a sealing device in the bottom of the casing. The hole is washed clean, wellhead controls are installed, and the well is allowed to flow. Sometimes the formation face is cleaned with a flush or mild acid solution.

Figure 5–10
Screens

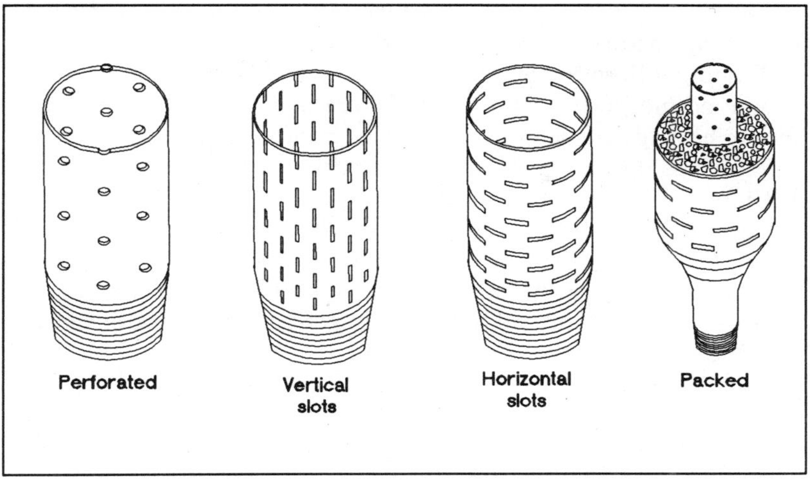

Gravel packing is combined with a screen sized for more severe sand conditions. Gravel is a common term in these cases because many wells in the past were packed with gravel. The sand size for gravel packing is selected by analyzing a sample of the formation sand. Then the precisely sized sand suspended in fluid is pumped into the hole on the outside of the liner screen. This restricts sand production by limiting the size of sand grains entering the wellbore. Smaller sized sand grains that enter through the sandpack flow out of the wellbore with produced fluids. Gravel packing is not common in horizontal wells and may need improved technology. Injecting special polymers into the formation is another procedure for restricting sand production that may have future application.

Screens and gravel packing have one major disadvantage. They freguently stick in the hole, requiring fishing for removal. This may or may not be difficult depending upon conditions, but generally it is a problem in high-angle and deviated holes.

Some horizontal wells are completed with a casing or a liner run to the bottom of the horizontal hole and cemented. This is similar to conventional completions and often used for higher pressure wells, those requiring acidizing and fracturing, and early discovery wells. It also is common for wells with multiple zones, especially when stimulating each zone separately. Sometimes zone isolation is improved by placing inflatable packers between each zone during casing and cementing. These completions generally apply to medium-turn and long-turn horizontal wells and sometimes to extended-reach wells. They may require larger hole and casing sizes.

The simplest completion in these cases is to perforate and allow the well to flow naturally. Perforating guns or jets are run on coiled or regular tubing. Some operators perforate on the low side to keep debris from entering the wellbore. It may be necessary to stimulate based on tests and other information described later in this section. The tubing is placed in the well, frequently set on a packer. The pressures are confined inside the tubing for higher pressured wells. The procedure should be completed by connecting the surface wellhead controls, installing production equipment, testing the well, and placing it on production (see Fig. 5–11).

Tubing and packer combinations are more common for higher pressure wells. The packer is run into the well with tubing and set with the same tubing. This confines high-pressure oil and gas inside the stronger tubing and provides a method of controlling pressures for reworking the well later if needed. A hydraulically set packer is run. Mechanically set packers require rotating the tubing, which can be difficult and sometimes impossible in high-angle or horizontal holes.

One alternative is to run a packer on a wireline if conditions permit and then run tubing and seat it in the packer. A permanent type completion should be used for very high-pressure wells. The packer and tubing combination is installed first and then perforating is done, usually with coiled tubing or special guns (see Fig. 5–12). This may be limited by the diameter of the regular tubing and length of the horizontal section.

Figure 5–11
Horizontal completions

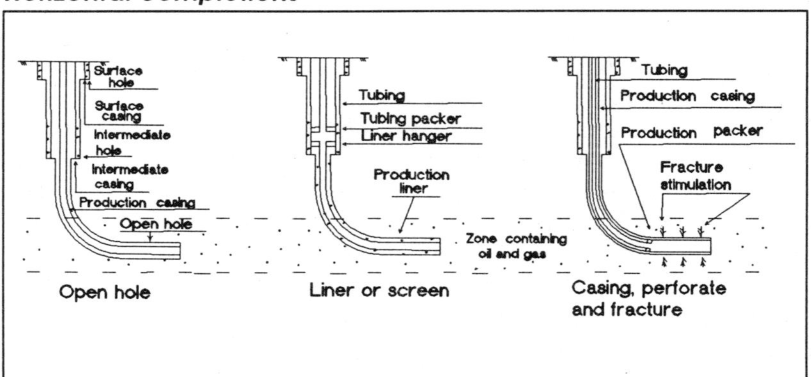

Figure 5–12
Special perforating guns
(courtesy of Halliburton)

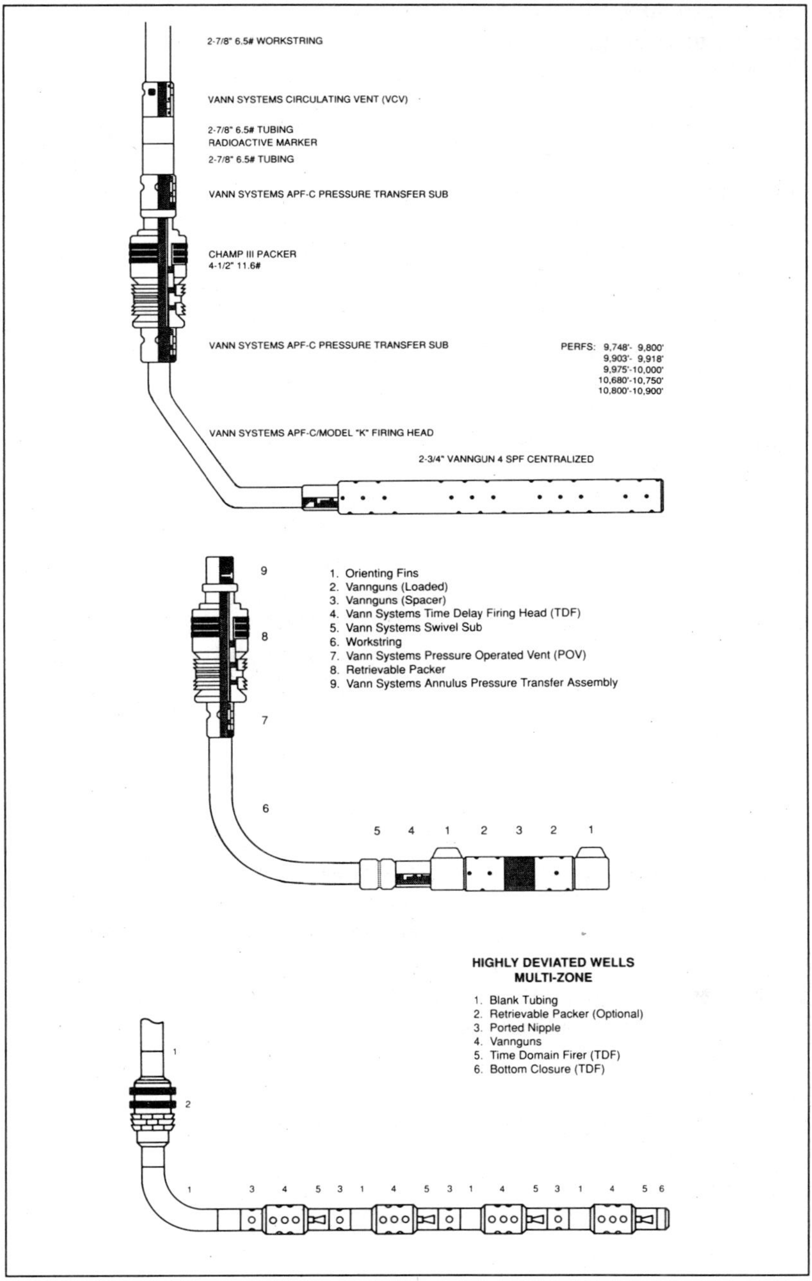

More complex completions, common in vertical and directional holes, are seldom if ever used. Two zones may be completed together by treating them as one and producing them comingled. Alternately, the lower zone may be completed and sealed with a retrievable plug or sliding sleeve, and then the upper zone may be completed. Then the plug is recovered or the sleeve is opened to produce both zones comingled. Dual completions are possible by producing the lower zone through the tubing and the upper zone through the tubing casing annulus. The procedure could be expanded to include three zones, but this is highly speculative.

These types of horizontal completions can become mechanically complicated. The risks of losing perforating tools and other tools increases. Recovering them by fishing is a high-risk procedure that is not always successful as previously described in the section about fishing. This is a good reason to use slotted or perforated liners and an open-hole completion whenever possible.

Most wells require stimulation during the completion. This ranges from a light flush for washing the face of the formation to a deep fracture treatment. If formation stimulation requirements are known, then the well is treated accordingly. Flowing and pressure buildup tests are used to determine if stimulation is needed and what the treatment requirements are. There are various programs for developing acidizing and fracturing treatments. Sometimes data are gathered for the programs with procedures such as microfracturing and strain relaxation.

Mild stimulations include flushing or washing the formation with chemically treated fluids. Deeper stimulation procedures include fracturing by injecting acid or sand-laden fluids (or both) into the formation. Small volumes are injected for shallow stimulations and large volumes for deeper stimulations. Some stimulations are very large, such as the injection of more than 8 million pounds of sand, suspended in a gel fluid, into a horizontal well in the North Sea.

Horizontal wells may be produced by artificial-lift methods, such as pumping, if they do not flow naturally. There are various pumping procedures but many have limitations in horizontal wells. Standard rod pumps may be placed in the vertical or low-angle section. However, this often is too high above the producing formation and restricts production. Placing the pump deeper in the curved or horizontal sections causes pump and rod wear and related problems. Hydraulic pumping systems should be used when applicable.

Submersible pumps with a combined centrifugal pump and electric motor are highly efficient under certain conditions, especially in high-productivity wells. The pump is connected to the tubing and lowered into the well with a trailing, insulated electrical power conduit fastened to the side of the tubing. The pump cannot pass through very short-radius curves without possible damage but can be run through curves with moderate turn radii. For operation, it is positioned in a straight section such as the vertical, tangent, or horizontal hole or sometimes in a section with a long radius.

Wells often have a productive life of 20 years or more. During this time, they require remedial work such as repairs, replacing pumps or other downhole equipment, cleanout, and additional stimulation. The wells should be reworked with completion or workover rigs. The original completion equipment should be removed (when required) in the reverse order of installation. Repair operations should be conducted in the same general manner as the completion procedures. Equipment and procedures are available for remedial work in horizontal and high-angle wells.

BIBLIOGRAPHY

B. S. Aadnoy and M. E. Chenevert. "Stability of Highly Inclined Boreholes." *SPE Drilling Engineering* (December 1987): 364–374.

American Petroleum Institute. *API Specification 10D, Special Edition.* API Specifications Casing Centralizers. Dallas: American Petroleum Institute, 1986.

R. Badry. "Production Logs Optimize Horizontal Tests." *World Oil* (March 1991): 59–66.

S. L. Barrett and R. Lyon. "Navigation Drilling Effective in Horizontal Wells in the Java Sea." *Drilling* (May/June 1988).

E. R. Blanco. "Horizontal Wells, Part 8—Hydraulic Fracturing Requires Extensive Disciplinary Interaction." *Oil & Gas Journal* (December 31, 1990): 112–117.

D. G. Calvert and D. K. Smith. "API Oilwell Cementing Practices." *Journal of Petroleum Technology* (November 1990): 1364–1373.

D. D. Clark and J. W Barth. "Calculator Programs Guide Directionally Drilled Wells through Tangled Thums Lease." *Oil & Gas Journal* (October 10, 1983): 90–112.

J. D. Clegg. "Rod Pumping Selection and Design." *Petroleum Engineer International* (July 1991): 44–48.

D. D. Cramer. "Stimulating Shales." *Oil & Gas Journal* Part 1. (April 22, 1991): 53–61; and Part 2. (April 29, 1991): 56–61.

W. L. Daniel and W. H. Fertl. "Logging High-Angle, Long-Reach Boreholes." *Oil & Gas Journal* (December 3, 1984): 103–108.

F. Daviau, et al. *Pressure Analysis for Horizontal Wells.* SPE 14251. Presented at the Society of Petroleum Engineers Annual Technical Conference and Exhibition. Las Vegas, NV, September 22–25, 1985.

H. Delafon. "BHA Prediction Software Improves Directional Drilling." *World Oil* Part 1. (March 1989): 45–50; and Part 2. (April 1989): 51–60.

D. Dennis and D. Jetelina. "New Logging Approach Detects Fractures in Horizontal Wells." *Petroleum Engineer International* (September 1990): 30–36.

W. Dickinson and R. W. Dickinson. *A Second Generation Horizontal Drilling System.* IADC/SPE 14804. International Association of Drilling Contractors/Society of Petroleum Engineers. Dallas, TX, February 10–12, 1986.

W. Dickinson and R. W. Dickinson. *Horizontal Radial Drilling System.* SPE 13949. Presented at the Society of Petroleum Engineers 1985 California Regional Meeting. Bakersfield, CA, March 27–28, 1985.

J. R. Duda, et al. "Pressure Analysis of an Unstimulated Horizontal Well with Type Curves." *Journal of Petroleum Technology* (August 1991): 988ff.

N. Eaton. *Coring the Horizontal Well.* American Society of Mechanical Engineers Drilling Technology Symposium. New Orleans, LA, 27 (January 1990): 65–69.

M. J. Economides, et al. "Performance and Stimulation of Horizontal Wells." *World Oil* Part 1. (June 1989): 41–45; and Part 2. (July 1989): 69–76.

H. B. Evans. "Evaluating Differences between Wireline and MWD Systems." *World Oil* (April 1991): 51–60.

W. H. Fertl and S. B. Nice. *Well Logging in Extended-Reach and Horizontal Wellbores.* OTC 5828. Offshore Technology Conference. Houston, TX, May 2–5, 1988.

J. D. Fultz and F. J. Pittard. "Bottomhole System Works Over, Completes Horizontal Wells." *World Oil* (March 1990): 48–50.

T. M. Gaynor. "Downhole Control of Deviation with Steerable Straight-Hole Turbodrills." *SPE Drilling Engineer* (March 1988): 50–56.

J. F. Greenip Jr. "How to Design Casing Strings for Horizontal Wells." *Petroleum Engineer International* (December 1989): 34–38.

D. A. Gust and R. R. MacDonald. "Rotation of a Long Liner in a Shallow Long Reach Well." *Journal of Petroleum Technology* (April 1989): 401–404.

J. Haney and G. Folmnsbee. "Coil Tubing Improves North Sea Squeeze Cementing." *Petroleum Engineer International* (August 1991): 28–34.

P. M. Hanson, et al. *Investigation of Barite "Sag" in Weighted Drilling Fluids in Highly Deviated Wells.* SPE 20423. Society of Petroleum Engineers. New Orleans, LA, September 23–26, 1990.

R. S. Hoch. "Cementing Techniques Used for High-Angle, *S*-type Directional Wells." *Oil & Gas Journal* (June 22, 1970): 88–97.

D. Holbert. "Conventional Tubulars Can Cut Cost of Drilling in Horizontal Holes." *Oil & Gas Journal* (June 3, 1985): 68–73.

M. R. Islam and A. E. George. "Sand Control in Horizontal Wells in Heavy-Oil Reservoirs." *Journal of Petroleum Technology* (July 1991): 844–853.

C. A. Johancsik, D. B. Friesen, and R. Dawson. "Torque and Drag in Directional Wells—Prediction and Measurement." *Journal of Petroleum Technology* (June 1984): 987–992.

M. M. Kamal. "Expected Developments in Transient Testing." *Journal of Petroleum Technology* (August 1991): 898–997.

S. C. Lien, et al. "The First Long-Term Horizontal-Well Test in the Troll Thin Oil Zone." *Journal of Petroleum Technology* (August 1991): 914–972.

J. L. McAlpine and S. D. Joshi. "Horizontal-Well Pilot Waterflood Tests Shallow, Abandoned Field." *Oil & Gas Journal* (August 5, 1991): 46–47.

R. Matson and R. Bennett. "Horizontal Wells, Part 7—Cementing Horizontal Holes Becoming More Common." *Oil & Gas Journal* (December 17, 1990): 40–46.

S. B. Nice and W. H. Fertl. "Logging, Completing Extended-Reach and Horizontal Wells." *World Oil* (March 1991): 49–56.

S. Ogden. "Inflatable Packers Provide Options for Horizontal Wells." *Petroleum Engineer International* (November 1991): 37–42.

Petroleum Engineer International. "Many MWD Choices Available." *Petroleum Engineer International* (May 1990): 37–40.

Petroleum Engineer International. "MWD Systems Expand Capabilities." *Petroleum Engineer International* (May 1991): 18–23.

P. E. Pilkington. "CBLs Can Evaluate Cement Integrity between Two Casing Strings." *Oil & Gas Journal* (December 10, 1990): 42–45.

F. J. Pittard and J. D. Fultz. *The Slimdril Horizontal Drilling System.* FOO508. PD—Vol. 27. Drilling Technology Symposium, the American Society of Mechanical Engineers, 1990.

F. L. Pruitt, K. C. Ross, and J. Woodruff. *Drilling with Steerable Motors in Large Diameter Holes.* SPE 17190. Society of Petroleum Engineers/International Association of Drilling Contractors Drilling Conference, Dallas, TX, February 1988.

L. Reiss. *Horizontal Well Production after Five Years.* SPE 14388. Presented at the SPE Annual Technical Conference and Exhibition. Las Vegas, NV, September 22–25, 1985.

R. H. Reiley, et al. *Cementing of Liners in Horizontal and High-Angle Wells at Prudhoe Bay Alaska.* SPE 16682. Society of Petroleum Engineers. Dallas, TX, September 27–30, 1987.

R. H. Reiley, et al. "Improved Liner Cementing in High-Angle/ Horizontal Wells." *World Oil* (July 1988): 69–74.

D. D. Sparlin and R. W. Hagen, Jr. "Gravel Packing Horizontal and High-Angle Wells." *World Oil* (March 1992): 45–49.

M. Taylor and N. Eaton. "Horizontal Wells, Part 5—Formation Evaluation Helps Cope with Lateral Heterogeneities." *Oil & Gas Journal* (November 19, 1990): 56–66.

M. Wasson, F. Pittard, and L. Robb. "Horizontal Workover with Coiled Tubing and Motors." *Petroleum Engineer International* (June 1991): 40–42.

M. B. Webster, G. E. Otott, Jr., and K. L. Rice. *Cementing High-Angle Wells Using Cement-Expanded Formation Packers and/or Casing Rotation.* SPE/IADC 16136. Society of Petroleum Engineers/ International Association of Drilling Contractors. New Orleans, LA, March 15–18, 1987.

C. White. "Horizontal Wells, Part 6—Formation Characteristics Dictate Completion Design." *Oil & Gas Journal* (December 3, 1990): 58–64.

J. P. Wilkirson, et al. "Horizontal Drilling Techniques at Prudhoe Bay, Alaska." *Journal of Petroleum Technology* (November 1988): 1445–51.

J. S. Williamson. *Casing Wear: the Effect of Contact Pressure.* SPE 10236. Society of Petroleum Engineers. San Antonio, TX, October 5–7, 1981.

D. J. Wilson and M. F. Barrilleaux. "Water-Packing Techniques Successful in Gravel Packing High-Angle Wells." *Oil & Gas Journal* (July 8, 1991): 30–37.

M. A. Wilson. "Cementing Horizontal Wells in Preparation for Stimulation." *World Oil* (October 1989): 72–84.

B. M. Zagalal and P. J. Murphy. "Reservoir Stimulation of Horizontal Wells in the Helder Field." *Journal of Petroleum Technology* (August 1991): 906–913.

C. Zimmerman and D. Winslow. "How to Select the Right Tools for Stimulating Horizontal Wells." *World Oil* (November 1989): 53–57.

INDEX

D

E

F

G

H

I

J

K

L

M

N

O

P

R

S

T

U

V

W

Y